A guide to
and culture of

Flowering Plants

and their seed
important to

Bobwhite Quail

Walter Rosene
John D. Freeman

Published by
Morris Communications Corporation

Library of Congress Cataloging-in-Publication Data

Rosene, Walter, 1912-
A guide to flowering plants and their seed important to bobwhite quail.

Bibliography: p.
Includes index.
1. Bobwhite--Food. 2. Bobwhite--Habitat. 3. Birds--Food. 4. Birds--Habitat. I. Freeman, John Daniel, 1941- . II. Title.
QL696.G27R68 1988 598'.617 88-31505
ISBN 0-9618270-2-5
ISBN 0-9618270-1-7 (pbk.)

Painting on cover and sketches by Richard A. Parks

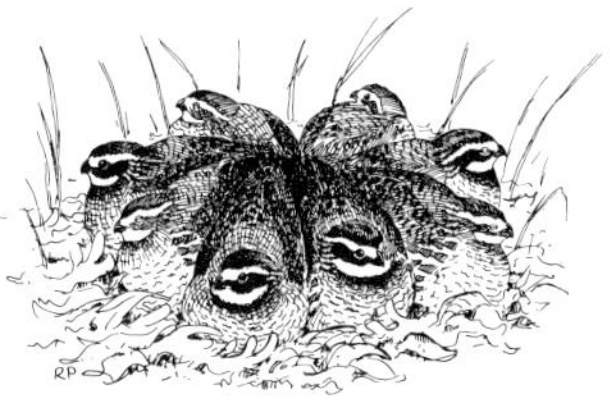

CONTENTS

FOREWORD

Quail have a large appetite and eat a wide variety of foods. Seed, insects and green plant material make up a substantial part of their diet, but during hunting season, they eat mostly seed.

Every quail hunter wants to know what the birds he shoots have been eating. Most of us will open the crop on the quail we kill and examine its contents, but we are not able to identify readily what these birds have been eating, with the exception of the obvious, such as acorns, corn, oats and other small grain.

So when most of us ask "What are quail eating?" the answer is seed – but, oh, what a wide variety. That's why this book is so important. It gives us a pocket-sized reference to the most prevalent food quail consume during hunting season.

But this book isn't just for the quail hunter. Because of its magnificent photographs and detailed descriptions of flowers, plants and seed, it appeals to a wide variety of readers, including nature lovers, gardeners, botanists, flower lovers and others.

Walter Rosene has devoted his life to the study of the bobwhite quail: its habitat, diet, protection and reproduction. He has one of this country's largest collections of seed from trees, plants and grasses, amassed since 1937 when he first became interested in seed collecting. He has almost instant recall in the field for both scientific and common names of almost any plant.

He has been examining crops of quail for longer and with more intensity than any individual in the United States. His book, *The Bobwhite Quail, Its Life and Management*, first was published by Rutgers Press in 1969. We at Morris Communications Corporation were proud to print its third edition – which is still in demand – in 1984.

It's been my pleasure to have worked with Walter both in the field and in the office on this new book, which represents the compilation, in print, of a lifetime of seed collecting.

I hope this book will be helpful to all who enjoy and love this greatest of game birds – the bobwhite quail.

Dr. John Freeman of the Department of Botany and Microbiology at Auburn University also has made a great contribution to this book. He was associated with Walter for about the last year and a half of the project. After Walter had made many attempts to find equipment to photograph seed, he found Dr. Freeman, who was able to photograph even the smallest seed, using microscopic equipment at Auburn. He also helped research and write some of the text.

W.S. Morris III
Augusta, Georgia
October 1988

INTRODUCTION

This manual is intended to serve as a guide to the kinds of plant species whose seed are consumed by bobwhite quail (*Colinus virginianus* L.) as well as other birds. It may also be useful to gardeners who may want to encourage wild plants as food and shelter to attract birds. Brief suggestions are made for culture of plants that will serve these ends, but our primary aim is to aid persons untrained in botany to identify quail food plants.

Many more or less technical references are available that enable plant identification by the use of keys. Other publications identify seed alone by keys or black and white photographs; a few identify plants or their seed or both by drawings (see references); but, to our knowledge, no field guide before this one has attempted to illustrate individual plant species by means of habit photographs, closeups of flowers, and seed in microscopic detail – all in full color.

The sequence followed in treating the plants in this book is similar to that in several other identification manuals and some state and regional floras. We chose, somewhat arbitrarily, to follow the same sequence of plant families and species as *Manual of the Vascular Flora of the Carolinas* (Radford, et al., 1968). A reader may wish to refer to that manual for detailed descriptions in some cases, particularly when efforts to identify some plant or seed fail to produce an exact match among the selection of species illustrated here. (Quail, after all, do eat seed of species not shown within these pages!) At least by following a "natural" system in our coverage, we have grouped the species together on the basis of fairly close genetic relationships. This generally means that plants with similar basic features and similar-appearing seed will be close together. This ought to make them easier to identify than if plants were arranged according to growth habit (trees, shrubs, herbs, vines, etc.) or flower color.

Descriptions and information about distribution and culture presented in this guide are very brief and relatively non-technical. Major features that identify particular species are noted. Seed formed by these species become available to quail and other birds usually from 2 weeks to 2 months after flowering, so fruiting times are not specified for each type of plant described. Illustrations herein should provide the information one needs to learn the identity of plants and seed important to quail.

The illustrations were chosen to show growth habit, floral features, and fruit/seed characteristics for each species. In a few cases, one photo will show both habit and flower, and in others the flowers are so inconspicuous that a closeup likely would not aid identification. Approximate magnifications or reductions are given for flowers. For example, (x 4) means the photo is 4 times the size of the actual flower and (x 1/4) means the photo is 1/4 the size of the actual flower. The scale illustrated with seed is always in millimeters (mm). A conversion scale for metric and English units is as follows: 1 inch = 25.4 mm. Thus, 1 mm = 1/25 inch; 5 mm = 1/5 inch; and 6 mm = 1/4 inch (approximate).

Birds have feeding habits peculiar to each species. Some species, like the purple martin, (*Progne subis* L.) feed entirely on insects, whereas others, like the bobwhite, feed on both seed and insects. Major shifts in the diet of quail occur twice a year. In winter they eat primarily seed and only a few insects. In spring they consume small amounts of plant material like young lespedeza leaves, but insects, which provide more protein than do seed, are eaten more and more as the season progresses. This shift toward insects as the primary food during summer is important for quail to go through the breeding cycle. Insects remain the main dietary item until fall when the shift is back to the food available in winter, namely seed. In summer and fall quail chicks require both insects and small seed.

Quail must feed every day of the year, and food must be available for them to do so.

Because of the truth of the above statement, we made no serious attempt to rank plants and seed as to relative importance to quail. Time of year, and both abundance and availability determine the kinds and amounts of seed consumed on a given day. A particular plant species may produce an abundance of seed, but if those seed fall upon a thick litter and filter into that mat of dead material, a ground-feeding bird likely would have difficulty finding them. Also, should seed remain on aerial stems of a plant for any prolonged period, they would be more likely to be eaten by a bird that feeds from perches rather than one on the ground like a quail.

To the sportsmen, hunters, wildlife specialists and managers, gardeners, and others who may use this guide, we pose a few other thoughts and suggestions. When you examine the contents of a crop (craw) from a bobwhite quail, is your first question usually, "What kind of seed is that?" And is your next question, "What does the plant look like that produces that seed?" If so, we hope you now will be able to answer these questions in many instances. Your observations of seed can be greatly improved by use of a small, inexpensive 10x folding magnifier which you should add to your hunting gear and always take into the field.

W.R. and J.D.F.

ACKNOWLEDGEMENTS

A very long list would be needed to acknowledge by name every person who assisted in one way or the other in the preparation of this guide. Many plantation owners and managers gave the senior author access to their lands to photograph plants and flowers and collect seed samples. Identification of certain questionable plant photos and seed was determined or verified by comparison with specimens in the Auburn University Herbarium, of which the junior author is curator. Some gaps in the senior author's seed collection that perhaps would otherwise have remained a frustration were filled with seed taken from voucher specimens in the Herbarium.

Certain individuals need special recognition for assistance beyond ordinary bounds and expectations. Dr. Alexander C. Martin, formerly of the U.S. Fish and Wildlife Service, a world authority on seed, now deceased, started the senior author's interest in seed collecting in 1937 and helped him often with seed identification in subsequent years.

Three persons with Morris Communications Corporation – Reginald B. Wells, Director of Corporate Communications; Andra Conley Eanes, Corporate Editor; and Lydia M. Inglett, Art Director – were more than generous with their time and efforts to aid us in completion of the guide.

Dr. Curt M. Peterson, Department of Botany and Microbiology, Auburn University, generously allowed us to use equipment in his lab to photograph seed, in particular the microscope, camera, etc., without which the task could not have been finished.

Joan V. Clements, secretary to the senior author, spent many hours indexing seed and typing the manuscript.

The authors are indebted to W.S. Morris III, Chairman and Chief Executive Officer, Morris Communications Corporation, for sponsoring the undertaking of this publication, encouraging us during preparation of the text and illustrations, and for insisting on highest levels of quality during the production of the printed material.

Finally, we must acknowledge any flaws or faults in fact regarding the plants shown or discussed within this text as being solely creditable to the authors. The expertise and standards of those whom we acknowledge here far exceed our own.

PINE FAMILY
Pinaceae

Pines *Pinus* spp.

Evergreen trees with 1 to 3 needles in a cluster. Some trees growing to a height of 100 feet. Species can be distinguished by needle length. Valuable as poles and lumber. Seed excellent quail food in fall and early winter.

Pine needles from top down:
- **Scrub Pine (*P. virginiana*) 2-6 cm. long**
- **Short-leaf (*P. echinata*) 8-12 cm. long**
- **Loblolly (*P. taeda*) 12-15 cm. long**
- **Long-leaf (*P. palustris*) 20-35 cm. long**

LONG-LEAF PINE *Pinus palustris*

PLANT

SEED

Description:
Large, slow-growing tree, reaching up to 70 feet in height from a long, deep taproot. Needles 15 inches long. Pollen cones shedding pollen in early spring. Seed cones 6 inches, releasing seed in fall.

Distribution:
Southern Coastal Plain and in isolated inland locations into northern Alabama.

Culture:
A valuable tree for poles and lumber found in well or poorly drained, sandy soils. Seed germinate readily and will grow if in contact with the soil. Natural reseeding occurs in open areas away from seed tree and may form dense stands. Seedlings 1 to 3 years of age are vulnerable to fire. After this period, burning is recommended in winter or early spring to control needle disease. After annual burning, growth is much more rapid. A burn in spring after terminal growth has begun is advisable if stand is too thick. This will kill some of the seedlings. Seed from this tree are desirable for quail; also eaten by other birds and rodents. Common ground cover beneath trees if soil has not been disturbed are various grasses (*Aristida stricta, Andropogon scoparius, A. tener*). Most grow in bunches and are commonly called "wire grass."

LOBLOLLY PINE *Pinus taeda*

PLANT

SEED

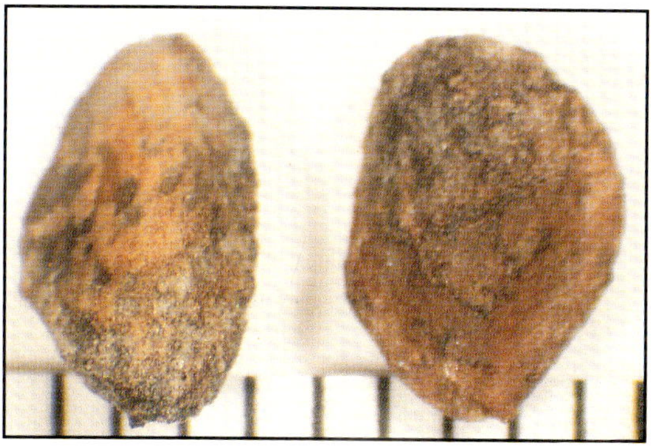

Description:
Large tree, growing to height of 60 feet. Needles in 3's, growing 6 to 9 inches long. Pollen cones shedding pollen in early spring.

Distribution:
Southern New Jersey to central Florida, west to southeastern Texas and southern Oklahoma, north to central Tennessee. Primarily Coastal Plain and Piedmont below 1,500 feet elevation.

Culture:
A very valuable timber tree. Volunteers in old fields after cultivation and on idle land. Stands can be increased in fall by natural reseeding. To accomplish this, burning or disking should be done in fall before seed drop and then by eliminating fire 8-10 years or until trees are large enough to tolerate fire. After this period, fire can be used to thin dense stands. Seed from this tree are an ideal quail food in fall and early winter.

SHORT-LEAF PINE *Pinus echinata*

BRANCH

SEED

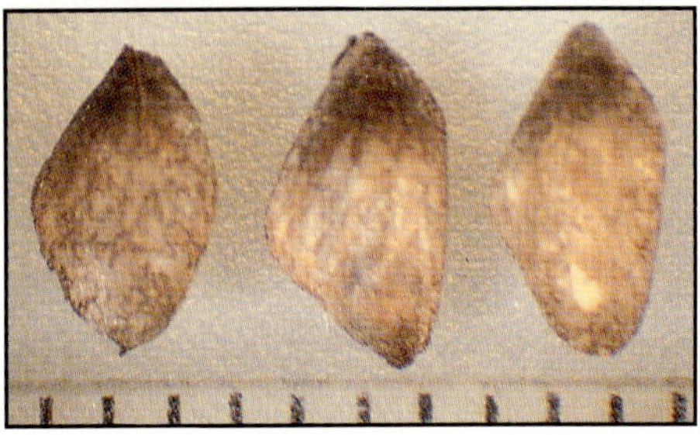

Description:
Large tree, growing to height of 60 feet. Needles in 2's, 3 to 5 inches long. Pollen shedding in early spring. Seed cones 1 to 3 inches long, dropping seed in early fall.

Distribution:
Eastern United States westward to Texas, but less common in high mountains and the Coastal Plain.

Culture:
Seeds heavily. Grows well on land formerly in cultivation and in woodlands where stands can be increased with the judicious use of fire. Valuable timber tree. Seed readily eaten by quail.

SCRUB PINE *Pinus virginiana*

BRANCH

SEED

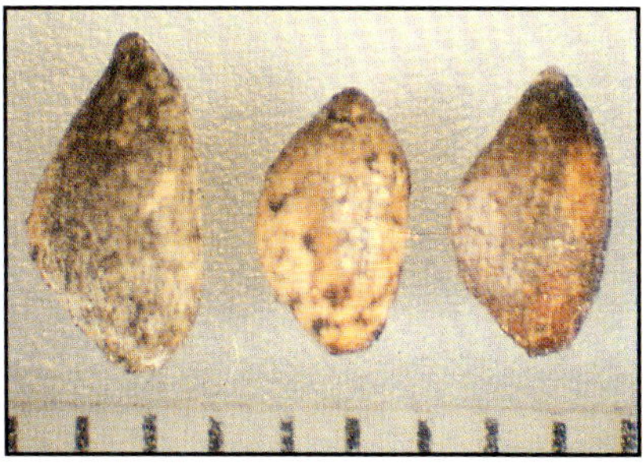

Description:
Small tree, growing to 30 feet tall. Needles in 2's, 1 to 2 inches long. Pollen shedding in early spring. Seed dropping in fall.

Distribution:
Inland portions of Atlantic and Gulf states mostly at higher elevations; rare in the Coastal Plain.

Culture:
Short height and many branches make this tree low in commercial value. Seeds heavily. Utilized by quail.

CYPRESS FAMILY
Cupressaceae

EASTERN RED CEDAR *Juniperus virginiana*

PLANT

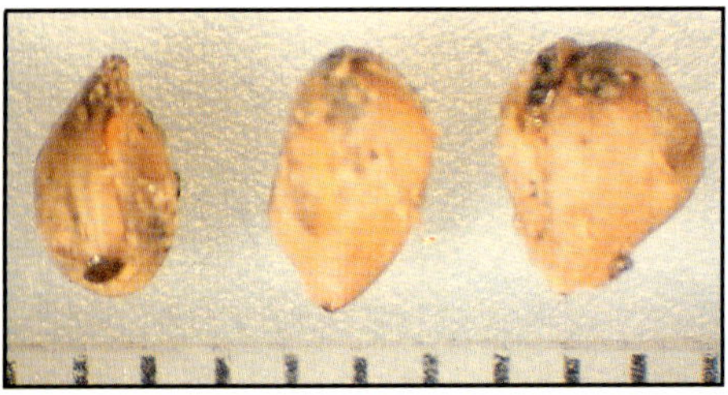

SEED

Description:
Medium-sized evergreen tree, growing in compact conical shape or more open. Leaves and twigs yellowish to bluish green. Male cones inconspicuous, at the ends of small twigs. Female cones borne on separate plants, waxy, bluish, resembling small berries, approximately ¼ inch in diameter. Blooming from January through March.

Distribution:
Canada to Georgia westward to Texas, Nebraska and Kansas.

Culture:
Grows on dry, rocky soils to fertile lands. Spread by birds along hedgerows, fences and brush piles. Common on limestone soils in Alabama and southern Tennessee where it is found in dense stands. Commonly used by quail and many species of song birds and mammals. Will not tolerate fire or disking. Used as a Christmas tree.

GRASS FAMILY

Poaceae

GRASSES

Grasses (with about 9,000 species) are economically the most important plant family; the cereal grains and most pasture and forage plants are grasses. Grass stems may be woody (in bamboos) or herbaceous. The herbaceous forms produce most of the grains eaten by birds and other animals. The following chart is offered to aid in separating grasses from sedges (*Cyperaceae*). Both families are important as quail food producers.

	GRASSES	SEDGES
Stems:	round in cross section and usually hollow	triangular (X.S.) and solid
Leaves:	flat or folded along midrib and arranged in 2 ranks (rows)	channeled along midrib and arranged in 3 ranks (rows)
Flowers:	often in flattened spikelets, except those with just 1 flower	often in cylindrical spikelets
Fruit:	grooved on 1 side (grain)	3-angled or lens-shaped (nutlet)

GOOSE GRASS *Eleusine indica*

PLANT

FLOWER (x 2)

SEED

Description:
Annual herb, growing erect up to 2 feet or prostrate. Root system dense. Flower heads with 2 to 5 finger-like projections, each with 2 rows of flowers on the lower side. Blooming from late spring to frost.

Distribution:
Only this introduced weedy species of *Eleusine* is found throughout the eastern United States. Has spread into various sections of the West.

Culture:
Common on soils of moderate to high fertility. As an annual it can withstand fire in winter and disking; abundant seed are present in the soil. A common weed in crops in warmer climates. The seed are used frequently in summer by quail chicks.

CROWFOOT GRASS *Dactyloctenium aegyptium*

PLANT

SEED

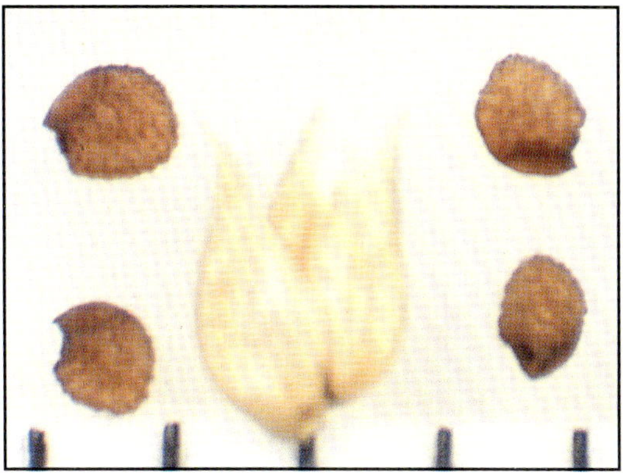

Description:
Usually an annual herb, growing erect to 2 feet or prostrate, but the root system sometimes over-wintering in the South. Root system becoming dense with maturation. Flower heads with 3 to 4 spreading spikes, each with a clawlike extension of the stem axis, hence the common name crowfoot.

Distribution:
Eastern United States. Introduced along the West Coast.

Culture:
A common weed found in crops on soil of average to high fertility. Can be burned in winter and disked; abundant seed are residual in the soil. Responds to regular crop fertilization. A desirable food for both young and adult quail in summer and fall.

FOXTAIL GRASS *Setaria geniculata*

PLANT

FLOWER (x 1/2)

SEED

Description:
Perennial herb, growing erect to 2 feet, or lower stems somewhat prostrate. Root system dense on mature plants. Seed (grain) produced in a dense spike-like terminal cluster, resembling a miniature fox tail. Blooming early summer to frost. About 8-10 species of *Setaria,* some perennials and some annuals, in the Southeast.

Distribution:
Eastern United States and in scattered locations to Texas.

Culture:
Grows in fields, along roadsides and in other waste places. Thrives in a wide variety of soil types. The annual species of foxtail can be increased by disking, and the perennials will withstand disking in alternate years. Uses residual fertilizer from cultivated crops. A quail food for young chicks and adults in summer and fall.

BULL or WATER GRASS *Paspalum boscianum*

PLANT

SEED

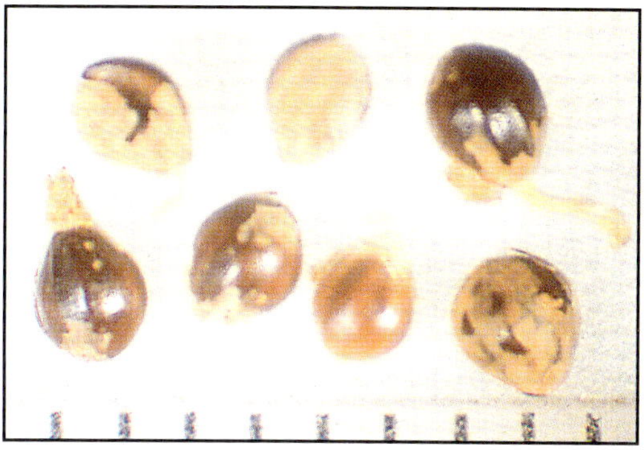

Description:
Annual herb, growing in tufts, up to 2 feet high. Mature plants with a dense root system. Flower head having from 1 to several spikes arranged alternately near the stem tip. Flowers on 4 rows on the underside of each spike. Seed (grain) brownish in color when mature. Blooming from summer to early fall. About 15 species of *Paspalum*, some annual and some perennial, in the Southeast.

Distribution:
Southeast, primarily in the Piedmont and Coastal Plain.

Culture:
Grows in damp places along drainageways and ditches. Will withstand fire and disking. Seed remain in the soil for several years. Will respond to fertilizer; however, it uses residual amounts from cultivated crops. A desirable quail food and also used by various other birds.

CRAB GRASS *Digitaria sanguinalis*

PLANT

SEED

Description:
Annual herb, growing along the ground or erect to 2 feet high and rooting at nodes. Root system compact. Flower head with 1 to 8 spikes, all originating from 1 point at the tip of the stem. Blooming from summer to frost.

Distribution:
Throughout the United States.

Culture:
Found in fields, along roadsides, waste places and lawns where it can be a pest. Grows in various soil types. Will withstand fire and disking. Uses residual fertilizer from crops and lawns, so it responds to applications of nitrogen. Commonly used as food by adult and young quail in summer and fall.

PANIC GRASSES *Panicum* spp.

PLANT

SEED

Description:
Annual and perennial herbs with erect growth habit varying from ground level to heights of several feet. Root systems from dense to open. Flower clusters from loose to compact. Seed (grain) flat on one side and round on the other, varying in size from a pin-head to a pea. Blooming from March to frost. *Panicum anceps*, the example shown above, has characteristics of many species. More than 150 species are recognized by some authorities on *Panicum*.

Distribution:
Throughout the United States.

Culture:
This genus has wide distribution, with plants growing in various soil types from dry to marsh land. The annuals can be encouraged by fire and disking. Most of the perennials will respond to light disking. Fertilizer requirements are broad. This genus is very important as quail food, particularly to chicks in summer.

BROWNTOP MILLET *Panicum fasciculatum*

PLANT

SEED

Description:
Annual herb, growing erect to 18 inches and branching from lower nodes. Root system well developed. Flowers on alternate spikelets. Older flowers and seed (grain) dark yellow to bronze. Blooming from late summer to frost.

Distribution:
Southeast, west to Arizona.

Culture:
Grown only in summer and best on moist soil of above-average fertility. An ideal seed source as food for quail chicks and adult birds when grown in strips of approximately 1/8th acre. Best results are in strips planted in June on recently cleared soil free of competition from other plants. Sow 30 lbs. of seed per acre and fertilize with nitrogen. Will mature in 6 weeks, so seed are available from September to December.

BROOMSEDGES *Andropogon* spp.

PLANT

SEED

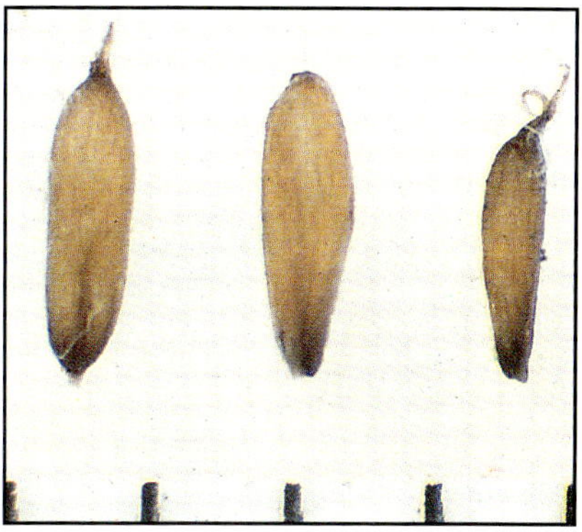

Description:
Perennial herbs, growing erect, 1½ to 3½ feet tall with leaves from crown of plant at ground level. Root system compact. Flowers inconspicuous. Seed (grain) with 2 scale-like attachments at base. Blooming in summer. Of about 150 species worldwide, only about 20 occur in the eastern United States. One of the most common is old-field broomsedge, *A. virginicus.*

Distribution:
Eastern United States.

Culture:
Found in open lands and in old fields on sandy or loamy soil. Will withstand fire. Heavy disking will destroy plants. Does not need fertilizer. It will volunteer in idle fields. Of no value as quail food. Ideal nesting cover when maintained in open stands so bare ground can be seen between plants. Two-year-old plants are ideal for nesting cover. Quail need unburned grass to build nests. Nests are found on edges of broomsedge left unburned on desirable sites, those edges around fields or adjacent to disked strips.

SEDGE *Cyperus* spp.

PLANT

SEED

Description:
Mostly perennial rhizomatous herbs. Stems erect to 2 feet tall. Flowers in heads of spikelets on short branches at tip of a larger triangular main stem. Seed heads tan to brown when mature. Blooming summer to fall.

Distribution:
Southeastern United States.

Culture:
Grows in various soil types; prefers moist land. Can be burned, not disked. Fertilizer requirements unknown.

SPIKE RUSH *Eleocharis obtusa*

PLANT

SEED

Description:
Annual herb, growing in tufts or clumps 18 inches high. Flower head dense at end of a straight stem. About 25 species of *Eleocharis* in the Southeast, including perennials as well as annuals.

Distribution:
Southeast.

Culture:
Grows in bogs, marshes, ditches, water holes and moist lands on wet soil. Effects of fire, disking and fertilizer unknown. Evidently quail find seed when they go to a watering hole.

BULRUSHES *Scirpus* spp.

PLANT

SEED

Description:
Annual or perennial herbs, growing in dense erect clusters to 6 feet tall. Root system compact. Flowers clustered in spikelets near the end of the stem. Blooming summer. There are approximately 25 species in the Southeast.

Distribution:
Throughout the United States.

Culture:
Grows on wet soil, fresh or alkaline. Will withstand fire. Disking in a dry period may decrease the stand. Seed of moderate use to quail. Seed commonly found on mud flats.

BEAK RUSHES *Rhynchospora* spp.

PLANT

SEED

Description:
Mostly perennial herbs, growing erect to 3 feet. Root system dense. For most species, flowers borne in loose or dense clusters near stem tip. Seed (nutlets) wrinkled with 1 to several bristles growing from base and a cap on terminal end. Blooming summer. More than 50 species occur in the eastern United States.

Distribution:
Eastern United States.

Culture:
Grows on partially drained soils of marshes and bogs. Effects of fire, disking and fertilizer unknown. Quail use is primarily in the Coastal Plain.

NUT RUSHES *Scleria* spp.

PLANT

SEED

Description:
Annual or perennial herbs, growing to 2 feet tall. Leaves along lower part of stems. Root system loose to compact. Long, triangular stem ending in a pointed bract prolonged above the clusters of flowers. Seed (nutlets) white, round, some wrinkled, some smooth, usually few per plant. Blooming summer to frost. Approximately a dozen species occur in the Southeast.

Distribution:
Eastern United States to Texas.

Culture:
Grows in open woodlands on dry to wet soil. Will withstand fire and light disking. Fertilizer requirements unknown. Much sought after by quail.

SEDGES *Carex* spp.

PLANT

SEED

Description:
Perennial herbs, growing erect to 4 feet tall. Root system compact. Flowers in clusters on stems near to and at the top. Seed (nutlets) tan to brown, enclosed in a sheathlike enclosure. Blooming summer to frost.

Distribution:
Throughout the United States.

Culture:
Grows in fields and open woods on damp soils. Will withstand fire. Heavy disking will destroy root crowns. Fertilizer requirements unknown. Moderate use by quail.

DAYFLOWER FAMILY
Commelinaceae

DAYFLOWER *Commelina erecta*

FLOWER (x 2)

SEED

Description:
Perennial herb with stems prostrate or erect up to 2 feet long. Root system thick. Flowers blue, lasting for just 1 day. Seed shape irregular, the surface often granular with long, narrow, darker strips. Flowering mid-summer.

Distribution:
Southeast.

Culture:
In dry, open woodlands on loamy to rocky soil. Will withstand fire and light disking. Fertilizer requirements unknown. Use by quail moderate.

RUSH FAMILY
Juncaceae

RUSHES *Juncus* spp.

PLANT

SEED

Description:
Herbaceous annuals and perennials, growing upright to 2 feet. Root systems compact. Flowers inconspicuous on branches of upper part of stem. Seed sharply pointed, numerous in each capsule. Blooming from summer to frost.

Distribution:
Southeast.

Culture:
Grows on moist soil in pastures, edges of swamps and ponds. Effects of fire, disking and fertilizer unknown. Seldom used by quail.

LILY FAMILY
Liliaceae

GREEN BRIER/CAT BRIER *Smilax bona-nox*

PLANT

SEED

Description:
Perennial evergreen vine. Stems sometimes growing to 30 feet in height but usually forming thickets near the ground. Flowers greenish, inconspicuous. Berries round, brown to black, each with 1 large reddish seed. Blooming summer.

Distribution:
Southeast, but more common in the Coastal Plain.

Culture:
Occurs in thickets, woods, idle fields and along fence rows on soils of low to high fertility. Will withstand fire, but continued disking will destroy this plant. Fertilizer requirements not known. Used by quail in periods of snow for protection and food.

BAYBERRY FAMILY
Myricaceae

WAX MYRTLE *Myrica cerifera*

PLANT

SEED

Description:
Evergreen shrub, growing upright to 30 feet. Flowers inconspicuous, forming clusters of small drupes borne on short spurs from 2-year-old twigs. Drupes round with a light blue to white waxy coating, persisting throughout winter. Blooming spring.

Distribution:
Southeastern United States, north to New Jersey and west to Texas.

Culture:
Grows in open woodlands and abandoned fields on poorly drained lands of damp, sandy soil. Attains highest growth in the Coastal Plain. Will withstand fire, but burning decreases seed crop. Moderate use by quail in the Coastal Plain.

BEECH FAMILY
Fagaceae

OAKS *Quercus* spp.

Evergreen and deciduous trees and shrubs, producing nuts (acorns) at least partially enclosed by a scaly cup around the base. Acorns, particularly the smaller ones, are an important quail food. The most important oaks are those that have narrow leaves; their acorns are usually less than ½ inch in diameter. These are classified generally as live oaks, water oaks, laurel oaks, and ground or dwarf oaks. About 3 dozen species of *Quercus* occur in the Southeast.

LIVE OAK *Quercus virginiana*

BRANCH

SEED

Description:
Evergreen tree growing to a height of 50 feet, usually with branches widely spreading. Root system shallow. Leaves dropping intermittently in spring as new ones develop. Acorns borne singly or in pairs on short stalks, dropping from September to March. Flowering in early spring.

Distribution:
Coastal Plain.

Culture:
Grows on sandy soil with moderate drainage. Fire will scar trees and kill smaller ones.

WATER OAK *Quercus nigra*

BRANCH

SEED

Description:
Semi-evergreen tree growing to height of 60 feet, usually with the crown well branched. Leaves deciduous in northern portions of range, essentially evergreen in the southern. Root system shallow. Acorns borne singly or rarely in pairs on very short stalks, dropping in fall and early winter. Flowering in early spring.

Distribution:
Southeastern United States.

Culture:
Found along edges of streams in rich bottom lands, rarely in dry uplands. Fire will scar trees. Important quail food in fall and early winter.

RUNNING OAK *Quercus pumila*

BRANCH

Description:
Deciduous, low-growing dwarf tree often growing in patches (or colonies) with stems reaching only 18 inches in height. Spreading by long underground stems. Acorns maturing the second year after flowering. Blooming in spring.

Distribution:
Coastal Plain.

Culture:
Grows on low, sandy soil. If this oak is burned annually, it will not produce acorns. Fertilizer requirements unknown. Acorns desirable for quail.

WILLOW OAK *Quercus phellos*

BRANCH

SEED

Description:
Deciduous tree growing to 60 feet tall, well branched. Root system shallow, compact. Acorns borne singly or in pairs on short stalks or attached directly to small leafy twigs, dropping in fall. Flowering in early spring.

Distribution:
Eastern United States west to Texas and Oklahoma.

Culture:
Found along stream banks of wet bottom lands and fertile soils of uplands. Fire will scar tree. Acorns used as quail food in fall and early winter.

LAUREL OAK *Quercus laurifolia*

BRANCH

SEED

Description:
Semi-evergreen tree growing to 60 feet, well branched. Root system compact. Acorns dropping fall and early winter. Flowering in spring.

Distribution:
Primarily Coastal Plain from North Carolina to Louisiana.

Culture:
Commonly occurs along streams and swamps, on wet soils. Fire not recommended. Acorns used as quail food fall and early winter.

ELM FAMILY
Ulmaceae

HACKBERRY *Celtis occidentalis*

BRANCH

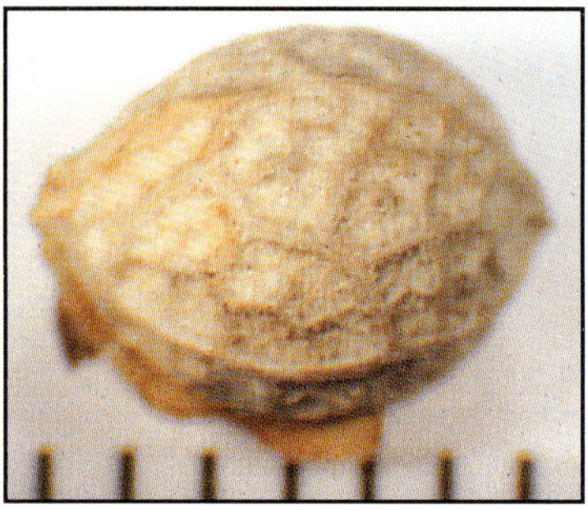

SEED

Description:
Large, deciduous tree, growing to 50 feet in height. Seed (drupes) maturing in fall. Flowering in May.

Distribution:
Southeastern United States.

Culture:
Grows in woodlands on soils of above average fertility. Fire will scar this tree. Fertilization not needed. Quail make moderate use of hackberry seed in winter.

MULBERRY FAMILY
Moraceae

MULBERRY *Morus rubra*

BRANCH

SEED

Description:
Moderate-sized, deciduous tree, growing erect to 40 feet. Fruit setting soon after blooming. Flowering in May and early June.

Distribution:
Eastern United States.

Culture:
Grows on rich, alluvial bottomland soils to altitudes of 2,000 feet. Fire will scar this tree. The fruits and seed probably used by quail more than is recorded; very few specimens have been collected in May and early June when mulberry is in fruit. However, at this time of year, quail diet is largely insects. Mulberry fruits are used extensively by other birds.

SMARTWEED FAMILY
Polygonaceae

SHEEP-SORREL *Rumex hastatulus*

PLANT

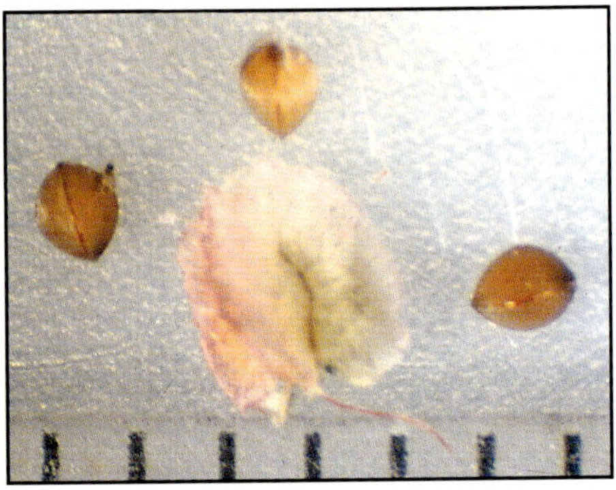

SEED

Description:
Also called sour-grass. Annual herb from a taproot, with angular, often 4-sided stems. Stems growing erect to 2 feet from a basal rosette of leaves at ground level. Flowers on individual plants either male or female. Entire plant with a pinkish color. Seed (nutlets) triangular, dark. Blooming early spring. Closely similar to *R. acetosella,* a rhizomatous perennial, with which often confused.

Distribution:
Common throughout Southeast.

Culture:
Grows in idle fields in various soil types. Disking will destroy plants. Fertilizer not needed. Seed used by quail for only a short period in late spring.

SMARTWEED *Polygonum pensylvanicum*

PLANT

FLOWER (x 1)

SEED

Description:
Annual herb. Stems often in thick stands with flower heads borne up to 2-3 feet. Flowers deep pink, rarely white. Seed (nutlets) dark brown to black. Blooming midsummer to frost.

Distribution:
Southeast.

Culture:
Low bottom lands on soils of moderate to high fertility. Will withstand flooding. Can be burned in winter. Disking in spring improves the stand. Responds to complete fertilizer. A preferred food for quail in winter.

PIGWEED FAMILY
Chenopodiaceae

PIGWEED *Chenopodium album*

PLANT

SEED

Description:
Also called lamb's-quarters. Annual herb from a taproot, the stems growing erect to 3 feet. Flowers tiny in branched, dense clusters at stem tips. Seed small, glossy black, maturing summer to fall. Blooming June to frost.

Distribution:
Throughout the United States except arid lands where it grows under irrigation.

Culture:
Common on fertile soils in cultivated fields, along field edges and in barnyards. Can be burned in winter. Disking improves the stand. Fertilizer not needed. Seed used moderately by quail in summer and winter.

AMARANTH FAMILY
Amaranthaceae

PIGWEED *Amaranthus hybridus*

PLANT

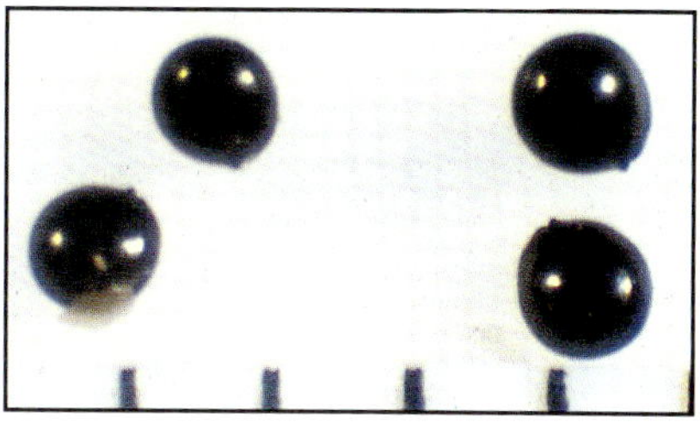

SEED

Description:
Annual herb from a taproot, the stem growing upright to 3 feet. Flowers greenish, small, in branched clusters at top of plant. Seed small. Blooming summer to frost.

Distribution:
Throughout the United States except arid sections where it grows under irrigation.

Culture:
Common in barnyards, idle lands and cultivated fields with soils of above average fertility. Can be burned and disked in winter. Does not need fertilizer. Moderate food value for quail.

POKEWEED FAMILY
Phytolaccaceae

POKEWEED *Phytolacca americiana*

PLANT

FLOWER (x 1/2)

SEED

Description:
Perennial herb from a large deep taproot, the stems growing to 8 feet. Flowers white or light green. Blooming summer to frost.

Distribution:
Eastern United States.

Culture:
Most common in windrows on land recently cultivated. Grows best on fertile soil. Will withstand fire and light disking. Fertilizer requirements unknown. Seed used moderately by quail in summer and early winter. Also used by other birds.

SEA PURSLANE FAMILY
Aizoaceae

CARPET-WEED *Mollugo verticillata*

FLOWER (x 3)

SEED

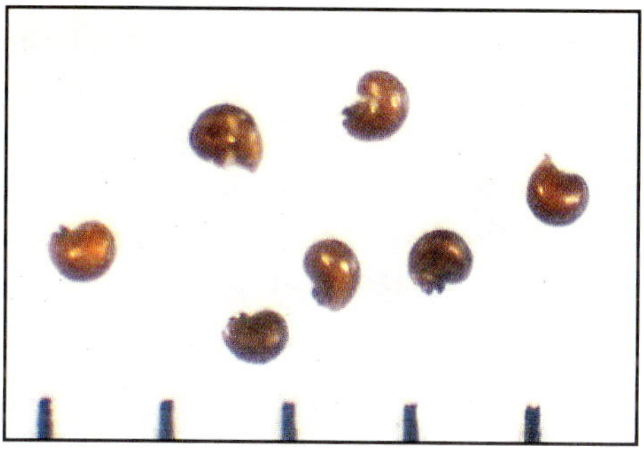

Description:
Annual, much branched, prostrate herb. Stems up to 18 inches long, rarely as much as 6 inches tall. Flowers small, the sepals green with white margin. Seed small, black and shiny. Blooming summer to frost.

Distribution:
Eastern United States.

Culture:
Grows in idle lands, primarily in fields that have been idle for 1 year. Soil requirements below average to good. Quail make moderate use of seed in summer and winter.

PINK FAMILY
Caryophyllaceae

COMMON CHICKWEED *Stellaria media*

PLANT

FLOWER (x 3)

SEED

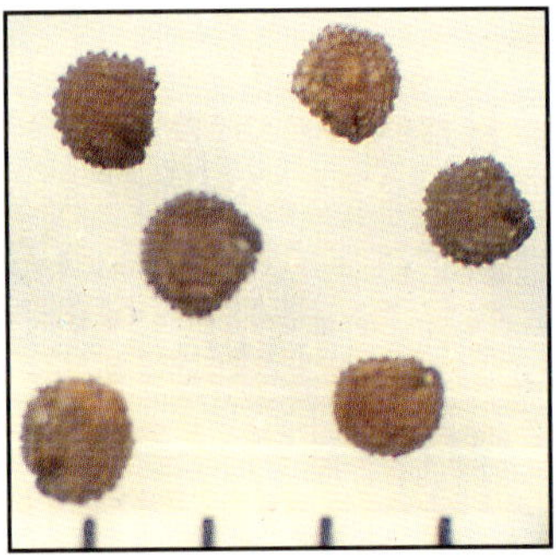

Description:
Small annual herb, with stems creeping along ground. Flowers small, white. Seed small, flatish, crinkled. Flowering spring and fall.

Distribution:
Eastern United States.

Culture:
Common in fields and idle lands with soil of average to good fertility. Can be disked and burned in winter. Fertilizer not needed. Seed of average use to quail in summer and fall.

MOUSE-EAR CHICKWEED *Cerastium glomeratum*

PLANT

FLOWER (x 3)

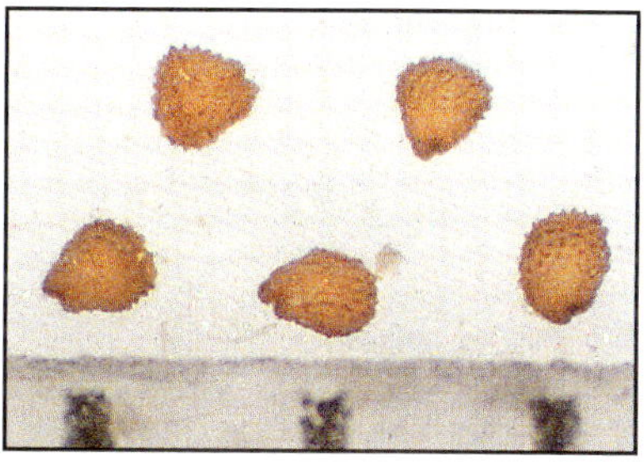

SEED

Description:
Winter annual, the stems erect to 6 inches high. Flowers small, white, numerous, clustered at the top. Seed small. Blooming March to May.

Distribution:
Throughout the United States except arid lands.

Culture:
Grows in gardens, fields, along roadways, in waste lands and recently idled lands of various soil types with moderate to good fertility. Light disking in late summer or fall will improve the stand the following year. Does not need fertilizer. Seed used in early summer by quail.

LAUREL FAMILY
Lauraceae

SASSAFRAS *Sassafras albidum*

PLANT

LEAF

SEED

Description:
Small tree or shrub, deciduous, growing to 20 feet. Stems aromatic, and root system well formed. Flowers yellow. Blooming in spring.

Distribution:
Eastern United States.

Culture:
Common in open woods, in idle fields, along roadways, fence rows and field edges. Grows in various soil types from poor to excellent fertility. Can be lightly burned in winter. Does not require fertilizer. Quail use sassafras as roosting cover, and the seed are of value for food in fall and winter.

MUSTARD FAMILY
Brassicaceae

PEPPERGRASS *Lepidium virginicum*

PLANT

BRANCH

SEED

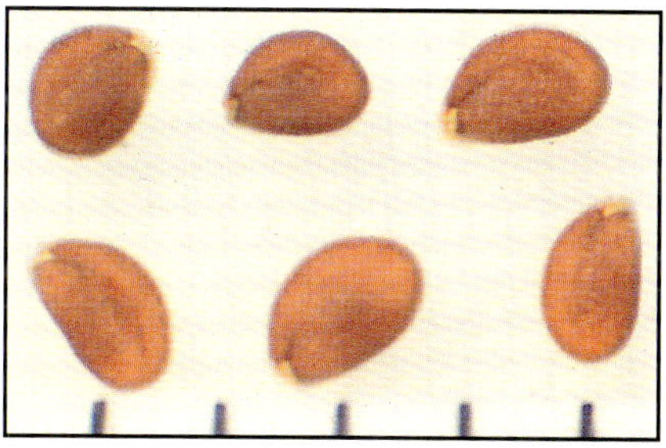

Description:
Also called poor-man's pepper. Freely branched annual herb. Stems upright to 18 inches. Petals white. Seed small, obovoid.

Distribution:
Southeastern United States.

Culture:
In fields, along roadways and in temporarily idle lands of various soil types. Seed of below average use by quail.

WITCH-HAZEL FAMILY
Hamamelidaceae

SWEET-GUM *Liquidambar styraciflua*

BRANCH

LEAF

SEED

Description:
Deciduous tree growing to 70 feet from a deep root system. Branches and twigs often with corky wings. Flowers small, in dense spherical clusters. Beaked capsules from female flower clusters forming a rough ball about 1½ inches in diameter, persistent on tree in winter.

Distribution:
Eastern United States.

Culture:
In woodlands from uplands to wetlands, of various soil types. Fire will scar older trees and kill sprouts up to 2 inches in diameter. This plant can become a pest because it reproduces rapidly in dense stands. Seed from older trees are extensively used as food by quail in fall and winter.

ROSE FAMILY
Rosaceae

BLACKBERRY *Rubus argutus*

PLANT

FRUIT (x 1 1/4)

SEED

Description:
Deciduous shrub. Branches angled, arched to a height of 4 feet, bearing many thorns. Root system loose. Flowers white. Ripe fruit an aggregate cluster of drupelets, the "blackberry." Blooming spring.

Distribution:
Southeast.

Culture:
In open areas in various soil types. Dense stands are formed in idle fields, along fence rows and field edges. The fruits are used by quail to some degree, and seed are eaten in fall and winter.

DEWBERRY *Rubus trivialis*

FLOWER (x 1 1/4)

SEED

Description:
Low-growing shrub with stems growing horizontally along ground, the flowering branches reaching heights of 6 inches. Stems with many thorns. Leaves persistent in winter. Flowers white, opening in early spring. Fruit an aggregate of drupelets, black. Other dewberries (*R. flagellaris, R. hispidus*) similar in appearance.

Distribution:
Southeastern United States.

Culture:
Volunteers in corn fields of previous year on sandy soil. Fire in spring will kill growth above ground. Disking in spring and continual cultivation will keep this plant under control. Of moderate use for quail.

CHEROKEE ROSE *Rosa laevigata*

PLANT

SEED

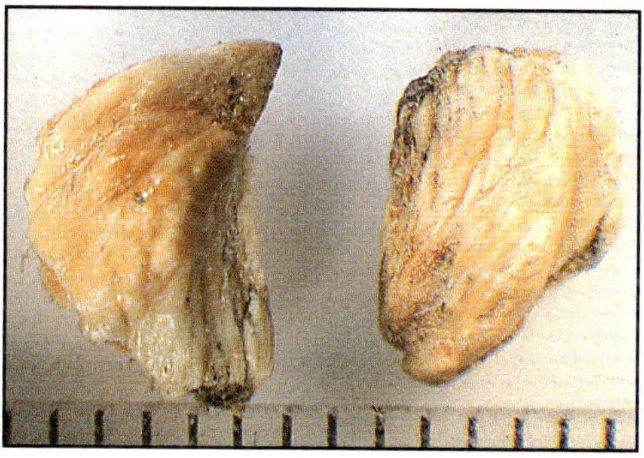

Description:
Evergreen vine-like shrub, climbing to 30 feet. Stems with many broad-based thorns. Flowers solitary, white. Blooming early spring.

Distribution:
Southeast, primarily the Coastal Plain.

Culture:
Along fence rows and woodland borders on soils of above average fertility. Will not withstand fire or disking. Of moderate use by quail.

MULTIFLORA ROSE *Rosa multiflora*

PLANT

FLOWER (x 3/4)

SEED

Description:
Evergreen shrub from a deep root system. Stems arching, branched, more or less upright, with many thorns. Flowers white in open terminal clusters. Blooming in May.

Distribution:
Southeast.

Culture:
Has spread along fence rows, in pastures and open woods in various soil types. A pest plant in many locations. Spreads rapidly by seed that are carried by birds. Fire in summer will kill plants. Seed (fruit) eaten by quail in winter. Hybrid tea roses are grafted onto this rootstock.

WILD CRABAPPLE *Malus angustifolia*

FLOWER (x 2/3)

SEED

Description:
Small, slow-growing, deciduous tree, reaching a height of 15 feet. Flowers pink, fading to white. Fruit round, green, about 1 inch in diameter, dropping in late summer.

Distribution:
Southeast, except southern Florida.

Culture:
Found in open woodlands and field edges. Protect from fire. The seed become available to quail as the fruits (crabapples) decay on the ground; thus may be eaten from fall into winter.

HAWTHORNS *Crataegus* spp.

PLANT

BRANCH

SEED

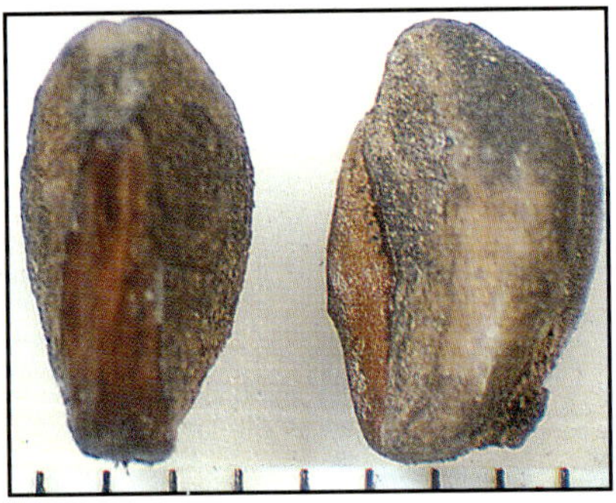

Description:
Small trees and shrubs, deciduous, growing erect to 20 feet. Branches of all species spiny. Flowers white to deep pink. Blooming spring to summer. About 15 species in the eastern United States.

Distribution:
Eastern United States.

Culture:
Three species found in dry locations, others along stream banks, low pastures and damp, open woodlands. Will not tolerate annual burning. Seed are used infrequently by quail.

WILD PLUM *Prunus angustifolia*

PLANT

BRANCH

SEED

Description:
Small, deciduous tree, growing erect to 10 feet, often forming thickets. Flowers white. Plums (drupes) red or yellow, ripening mid-summer. Blooming early spring.

Distribution:
Southeastern United States.

Culture:
Grows in old fields, open woods, idle areas with various soil types but seems to do best in sandy land. Ideal roosting cover for quail both in summer and winter because the thick plants shade out vegetation at the ground. Seed eaten by quail in late fall and winter.

WILD BLACK CHERRY *Prunus serotina*

BRANCH

BRANCH

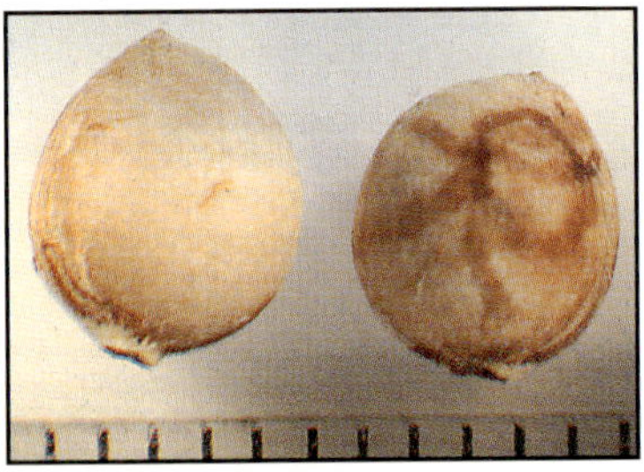

SEED

Description:
Large, deciduous tree growing to height of 70 feet from a deep root system. Flowers white, in cylindrical clusters. Cherries (drupes) maturing in July and August. Blooming spring.

Distribution:
Eastern United States.

Culture:
Grows in woodlands in various soil types but mainly in rich, moist locations. Fire will scar this tree. The fruits ripen in summer and drop to the ground so are thereafter available to quail. Seed used quite often by quail in fall and winter. The ripe fruits are much sought after by deer.

BEAN FAMILY
Fabaceae

LEGUMES

The Bean Family includes about 13,000 species, and ranks third in size worldwide behind the orchids (*Orchidaceae*) and asters (*Asteraceae*). The seed (beans, peas, etc.) are rich in protein as well as starch and oils, and many have been domesticated to serve as food plants for man and forage for animals. A large number of species form seed eaten by birds, including quail. The fruits or pods may contain 1 to many seed; the fruit is a legume technically if it splits lengthwise to release seed, but some non-splitting types are only 1-seeded (i.e., *Lespedeza*), and other fruits simply break apart between the seed (i.e., *Desmodium*). Flowers of members of the Bean Family vary greatly in form and arrangement on the plant, but most are showy. Typically there are 5 green sepals, sometimes fused into a tube, and 5 colorful petals of which 1 is considerably larger or smaller than the others.

SENSITIVE BRIER *Schrankia microphylla*

FLOWER
(x 3)

SEED

Description:
Perennial herb, with numerous small prickles. Stems often a yard long, growing close to the ground. Leaflets folding up when touched, on cloudy days and at night. Flowers tiny, forming a pinkish ball and attracting insects. Blooming from early summer to frost.

Distribution:
Southeastern United States.

Culture:
Commonly found at field edges, along roadsides and in open fields. Light disking encourages new plants, but heavy disking will destroy older ones. Grows in woodlands that are burned annually.

REDBUD *Cercis canadensis*

BRANCH

FLOWER (x 2)

SEED

Description:
Small tree or large shrub, commonly growing to a height of 15 to 30 feet. Flowers deep to pale pink, appearing on old stems in the spring before leaves come out. A very showy small tree commonly used as an ornamental.

Distribution:
Southeast, primarily on the clay soils of the Piedmont.

Culture:
Common in mixed stands of pines and hardwoods, usually on fertile soils. A prolific seeder, this small tree is rarely found in a dense stand. Seedlings will not tolerate fire, and older growth is damaged by excessive burning.

PARTRIDGE PEA *Cassia fasciculata*

PLANT

FLOWER (x 3/4)

SEED

Description:
Annual herb, the larger of the 2 common partridge pea species. Stems reaching a height of 3 feet. Flower yellow, very showy. Blooming June to September.

Distribution:
Southeastern United States.

Culture:
Common in open woodlands, idle fields, along roadways. Seed are very hard and remain viable in the soil for many years. Disking and burning are important to improve germination and to maintain stands. Plants may reappear after fire and land disturbance. Bees improve pollination and seed set. Beekeepers often locate hives in dense stands of partridge pea, and honey of very good quality is produced.

SMALL PARTRIDGE PEA *Cassia nictitans*

FLOWER (x 3)

SEED

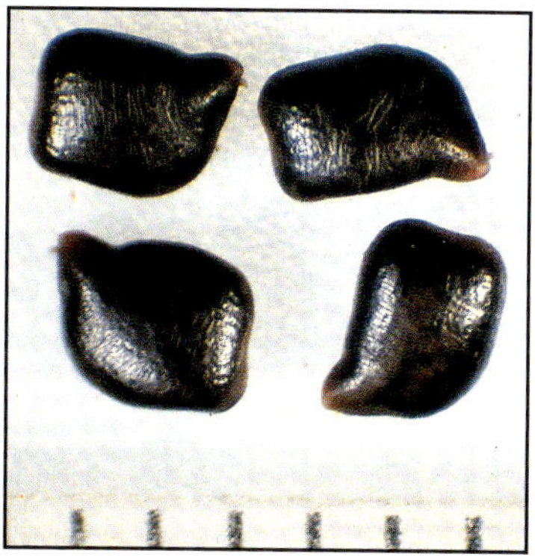

Description:
Also called wild sensitive plant. Annual herb, with a fleshy root system. The smaller of the 2 common partridge peas, growing up to 2 feet high. Flowers yellow, about ½ inch across. Much smaller than the large partridge pea (*C. fasciculata*) with flowers about 1 inch across. Closely similar to the sandhill partridge pea (*C. aspera*) of the Coastal Plain from Florida to South Carolina. Blooming from mid-summer to frost.

Distribution:
Southeastern United States.

Culture:
Grows in open woodlands and along roadways. This plant responds well to both disking and burning. To get volunteer stands over several years, land should be disked. Fire also scarifies the seed, improving germination. Seed are very hard, so they deteriorate slowly and remain viable in the soil for several years. A new stand may appear where none had been present for many years.

CRIMSON CLOVER *Trifolium incarnatum*

FLOWER
(x 1)

SEED

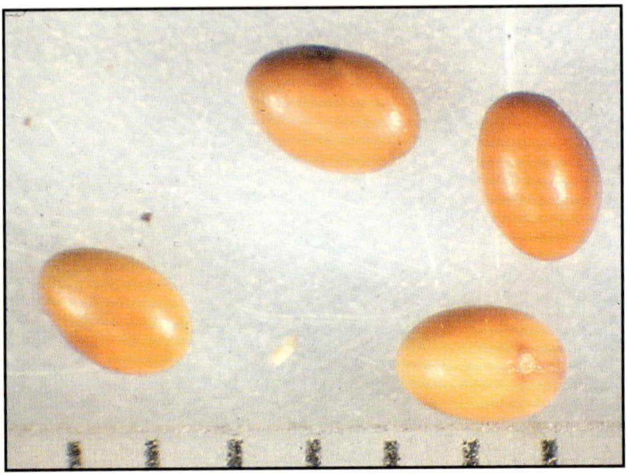

Description:
Annual herb. Flowers deep red, very showy. Stems growing from 1 to 2½ feet tall. Blooming from early spring to early summer.

Distribution:
Grows best in the lower portions of the Southeast but also is found elsewhere.

Culture:
Responds well to lime, phosphate and potash. Seeding rate 15-20 lbs. per acre with an inoculant. Widely used as a cover crop in fields and on roadsides. A prolific seeder.

RED CLOVER *Trifolium pratense*

FLOWER
(x 1)

SEED

Description:
Perennial herb. Stems upright, sometimes reaching 3 feet high, 1 or more stems from 1 root. Petals deep pink. Blooming from May to September.

Distribution:
Southeastern United States.

Culture:
Grows in open fields and along roadways. A desirable nectar source for honey. Responds to an application of lime, phosphate and potash.

WHITE SWEET CLOVER *Melilotus alba*

FLOWER (x 1 1/2)

SEED

Description:
Biennial herb. Growing up to 5 feet high. A large, loosely branched plant with white flowers. Blooming from May to September.

Distribution:
Eastern and central United States. Not generally common in the Southeast but may grow in dense stands in certain areas.

Culture:
Found in the open along roadsides, particularly on soils with high lime content, such as the Black Belt of Alabama and Mississippi. Where grown as a crop, responds well to applications of lime, phosphate and potash. Burning and disking not recommended.

YELLOW SWEET CLOVER *Melilotus officinalis*

PLANT

SEED

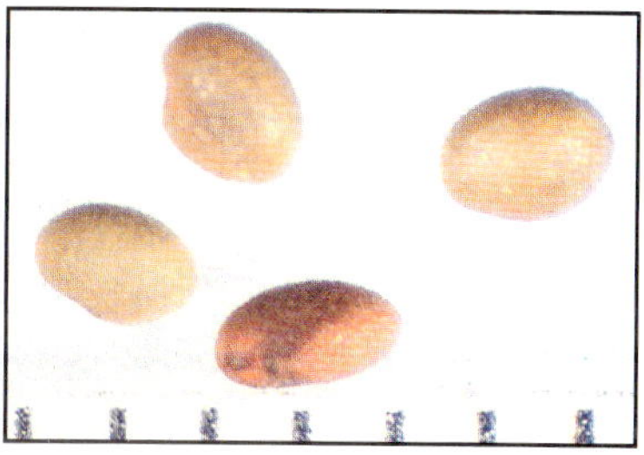

Description:
Similar to white sweet clover, except flowers yellow.

Distribution:
Same as white sweet clover.

Culture:
Same as white sweet clover.

INDIGO BUSH *Amorpha fruticosa*

PLANT

SEED

Description:
Shrub, growing 2 to 6 feet high. Flowers dark purple, usually on long stalks from the tips of branches. Leaves with as many as 20 opposite leaflets. Blooming April-June.

Distribution:
Southeast, primarily on upland soils with high lime content, often near streams.

Culture:
Avoid fire and disking. Fertilizer requirements unknown. This shrub will persist for several years.

SAMSON SNAKEROOT *Psoralea psoralioides*

PLANT

FLOWER (x 2)

SEED

Description:
Perennial herb from a long taproot. Stems growing up to 2½ to 3 feet tall. Flowers several, borne on leafless stalks 2-6 inches long. Petals blue to purple. Blooming in early summer.

Distribution:
Southeastern United States.

Culture:
Grows in open woodlands and fields. Can withstand fire and light disking. Fertilizer requirements unknown.

PENCIL FLOWER *Stylosanthes biflora*

FLOWER (x 3)

SEED

Description:
Low-growing perennial herb. Small yellow flowers. Usually growing in combination with grasses and other legumes; these other plants tending to make the pencil flower more erect. Blooming June-August.

Distribution:
Southeastern United States.

Culture:
Grows in open woodlands and field edges on dry, sandy soil. Will not stand heavy disking. Fertilizer requirements unknown.

BEGGARWEEDS *Desmodium* spp.

There are 23 species of beggarweed with different growth habits and various-colored flowers. Some grow erect, while others are prostrate. All native beggarweeds are perennial herbs. The Florida beggarweed (*Desmodium tortuosum*) is an annual herb.

BEGGARWEED *Desmodium nudiflorum*

PLANT

FLOWER (x 4)

SEED

Description:
Perennial herb, with a fleshy root system. Growing erect, from 1½ to 5 feet tall. Stems of 2 forms, 1 bearing leaves and the other flowers. Flowers pink, numerous, in a large, loose cluster. Blooming from June to October.

Distribution:
Southeastern United States.

Culture:
Commonly found in open woodlands with a mixed stand of pine and hardwood. Fire does not damage this plant because new growth comes each year from the root crown. A light disking will not damage this plant. If it is disked deeply, the root system will be destroyed. Responds to a light application of phosphate.

BEGGARWEED *Desmodium rotundifolium*

FLOWER (x 3)

SEED

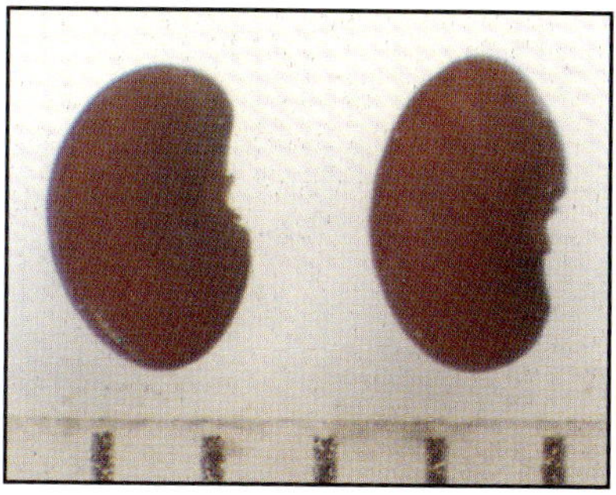

Description:
Also called dollar weed. Perennial herb, with a fleshy root system. Stems may be 3 feet long, close to the ground.

Distribution:
Southeastern United States.

Culture:
Responds to light disking, and fire will not damage the root system.

FLORIDA BEGGARWEED *Desmodium tortuosum*

PLANT

FLOWER (x 3)

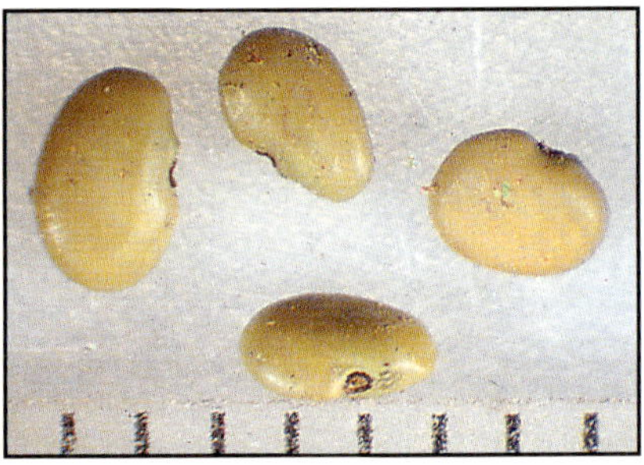

SEED

Description:
Annual herb, with tall stems growing upright. Flowers small, rather inconspicuous. Blooming from June to frost.

Distribution:
Southeastern United States, primarily in the sandy fields of the Coastal Plain.

Culture:
The best way to ensure a stand of Florida beggarweed is to sow seed in crop fields just before that crop is laid by. Seed germinate rapidly and will produce by the first fall. Establishment in this way eliminates the need for special application of fertilizer because this planting will get fertilizer that has been applied to the agricultural crop. When grown by itself, this plant responds to an application of phosphate and potash, and on sandy lands a light application of nitrogen is helpful. This plant was introduced in the United States years ago.

LESPEDEZAS *Lespedeza* spp.

There are 15 species of lespedeza: 2 annuals, 11 perennials and 2 shrubs. The annuals and the shrub species are introduced. These 4 species grow in thick stands, while the native lespedezas occur singly or in more open stands.

COMMON LESPEDEZA *Lespedeza striata*

PLANT

FLOWER (x 2)

SEED

Description:
Annual herb, the typical variety (common lespedeza) growing relatively close to the ground. The "Kobe" variety (developed from common) growing more upright and sometimes reaching a height of 18 inches. "Kobe" more available on the market, thus now used more than common. Both plants with rather inconspicuous flowers, blooming from June to September.

Distribution:
Particularly in lower portions of the Southeast where planted widely and commonly escaped from cultivation.

Culture:
Plant in February to March 15th in open woodlands, fields or small food plantings at the rate of 20-30 lbs. per acre and fertilize with 400 lbs. of 0-14-14 per acre. This plant will reseed for several years with proper maintenance. It is advisable to protect plantings from fire with a fire break because seed germinate early in March when most burning is done. Light disking and an application of phosphate are necessary to ensure a stand of plants in old plantings.

PROSTRATE LESPEDEZA *Lespedeza procumbens*

PLANT

FLOWER (x 2)

SEED

Description:
Perennial herb, with stems growing close to the ground. Petals reddish to purple. Blooming from June to October.

Distribution:
Southeastern United States.

Culture:
Commonly found in open woodlands, along field edges and roadsides. Must be disked lightly every 3 to 4 years to reduce competition from other plants. Fertilizer requirements unknown.

BICOLOR LESPEDEZA *Lespedeza bicolor*

PLANT

FLOWER (x 2)

SEED

BICOLOR LESPEDEZA *Lespedeza bicolor*

Description:
Shrub, growing to a height of 9 feet. Large, loose flowering head with many purplish flowers. Blooming from June to October.

Distribution:
Southeastern United States on sandy and clay soils.

Culture:
Land preparation is necessary when establishing a planting with 1-year-old seedlings. Size of a planting should be approximately 1/8th acre, 6 rows wide and 250 to 300 feet long. Rows should be 36 inches apart and plants set 2 feet apart in the row. Fertilizer can be applied either before or after planting using phosphate and potash at the rate of 500 lbs. per acre. No nitrogen. Plant in February and early March in well settled soil. It is recommended that when 1/8th acre strips are to be planted with seed that these strips be located on land relatively free of residual weed seed. This means that they be on land that has not been in cultivation for past several years. Open areas in woodlands are ideal locations for establishing strips with seed. The soil should be disked lightly and allowed to settle so that seed can be sown on a firm seed bed after a light disking. Cover very lightly. Sow seed in May at the rate of 1 lb. per 1/8th acre. Fertilizer rate same as for 1-year-old seedlings. Bicolor will require some maintenance. It is important that seed from this plant fall on ground that is relatively bare. An ideal planting will have only leaves of this shrub on the ground. This makes seed available to quail. When growth becomes short on the plants, it is recommended they be cut back, fertilized and disked. Disking will destroy weed growth in the planting but will not damage the bicolor crowns.
Deer will browse new and old plantings of bicolor. These animals are fond of new growth and have been known to kill entire plantings. Where deer are a potential problem, it is recommended that *Lespedeza thunburgii* be used.

THUNBURG LESPEDEZA *Lespedeza thunburgii*

PLANT

FLOWER (x 2)

SEED

Description:
Growth habits similar to bicolor lespedeza.

Distribution:
Grows well on the sandy clay soils of the southeastern United States.

Culture:
Similar to bicolor. This plant has recently been introduced and is relatively deer-resistant. Its seed are totally black and larger than bicolor and are eaten by quail as readily as bicolor. Bicolor and thunburg are also used by various species of songbirds, native and migratory.

JAPONICA LESPEDEZA *Lespedeza japonica*

PLANT

BRANCH

SEED

Description:
Shrub with growth habit similar to bicolor lespedeza, but somewhat smaller, reaching about 4 feet in height. Flowering head large, loose. Seed setting in July and August, dropping in September. Blooming late May, early June.

Distribution:
Now planted and grown on sandy clay soils in the Southeast.

Culture:
Similar to bicolor, but more deer-resistant than either bicolor or thunburg lespedeza. Seed difficult to combine because they mature and drop to the ground while the plant remains green. Because of the shorter stems, plantings of this species are easier to hunt in than the other shrubby lespedezas.

WILD SERICEA LESPEDEZA *Lespedeza virginica*

PLANT

FLOWER (x 2)

SEED

Description:
Perennial herb, with a fleshy root system. Stems growing erect, up to 3 feet. Flowers in clusters. Blooming from June to October.

Distribution:
Southeastern United States.

Culture:
Most commonly found growing in association with open stands of broomsedge. Can be burned annually with no damage to the plant. Light disking will not destroy the root system. Fertilizer requirements unknown.

WAND LESPEDEZA *Lespedeza intermedia*

PLANT

FLOWER (x 2)

SEED

Description:
Perennial herb, growing erect to a height of 2 to 3 feet in light, dry soils. Flowers purple. Blooming from mid-summer to October.

Distribution:
Southeastern United States.

Culture:
Commonly found in open woodlands, more or less as solitary plants. Never found in dense stands. Will tolerate light disking and fire. Fertilizer requirements unknown.

SERICEA LESPEDEZA *Lespedeza cuneata*

FLOWER (x 2 1/2)

SEED

Description:
Perennial herb, with a deep root system. Growing in a thick stand up to 3 feet tall. Blooming from June to October.

Distribution:
Southeastern United States.

Culture:
Seed are seldom eaten by quail. It was planted extensively for erosion control and hay in the 1930s and 1940s with a possible use for quail. More recently used along interstate highways to vegetate rights-of-way after construction. In the North quail use this plant for roosting because it offers protection from the cold and deep snow. Not recommended for use in the South. In the North, sow 30 lbs. seed per acre in a well prepared seed bed and fertilize with a mixture of 50 lbs. of phosphate and potash.

HAIRY LESPEDEZA *Lespedeza hirta*

FLOWER (x 2 1/2)

SEED

Description:
Perennial herb, with a fleshy root system. Stems very hairy, growing to a height of 2 to 3½ feet. Flowers and pods near the tip of each stem. Blooming from mid-summer to October.

Distribution:
Southeastern United States.

Culture:
Can withstand light disking and fire.

ROUNDHEAD LESPEDEZA *Lespedeza capitata*

PLANT

SEED

Description:
Perennial herb, with a large taproot. Several stems from 1 root, growing up to 3 feet tall. Flowers in bunches near the top of each stem. Seed remaining on plant well into winter. Blooming from mid-summer to October.

Distribution:
Southeastern United States.

Culture:
Found in open woodlands and dry fields. Does not grow in a dense stand. Tolerates fire and light disking.

SESBANIA *Sesbania macrocarpa*

PLANT

FLOWER (x 2 1/2)

SEED

Description:
Annual herb, growing from 2 to 10 feet tall, depending on fertility of soil. Flowers showy, yellow. Produces many seed in a long, black pod. Blooming late summer.

Distribution:
Southeastern United States.

Culture:
Grows well on poorly drained soils high in lime content. Should be planted in open fields. May be sown at the rate of 15 lbs. per acre. A dense stand of sesbania will shade out competing vegetation, so the seed will fall on bare soil and be available to quail. This plant fits in well where soil drainage is not suitable for other legumes. Should be fertilized with 200 lbs. of phosphate per acre. In a few instances, insect damage has been observed in thick stands. Sesbania will volunteer for several years with annual disking in early summer.

BLACK LOCUST *Robinia pseudo-acacia*

BRANCH

SEED

Description:
Deciduous tree, growing up to 50 feet tall. Leaves pinnate, with 7-19 leaflets. Flowers white, in large, drooping clusters. Seed maturing in fall. Blooming in May.

Distribution:
Central portions of Georgia, Alabama, and Mississippi northward in the eastern United States, especially in limestone areas.

Culture:
Grows in moist bottomlands and upland forests. Fire will damage trees of all ages. Valued for making fence posts, for which it was planted and grown widely in the past.

WILD INDIGO *Indigofera caroliniana*

PLANT

FLOWER (x 2 1/2)

SEED

Description:
Somewhat woody herb from a deep woody taproot. Stem erect or spreading, up to 4 feet in height. Leaves pinnate, with 7-15 leaflets. Flower cluster loose, the flowers opening over a long span of time. Blooming June-August.

Distribution:
Coastal Plain, southeastern United States.

Culture:
Grows in open woodlands, idle fields, along roadsides, in sandy soil. Will withstand light burning, but disking will kill the root system.

GOAT'S RUE *Tephrosia virginiana*

PLANT

FLOWER (x 3/4)

SEED

Description:
Perennial herb, from a root crown. Several stems from each crown, growing erect from 1 to 3 feet tall. Each stem with an abundance of very showy flowers. Each flower mixed red and white. Blooming May to July.

Distribution:
Southeastern United States.

Culture:
Found growing in open stands of pine with a few hardwood trees. Plants may occur singly or as a dense stand. Will tolerate light disking and burning. Fertilizer requirements unknown. This plant grows well on sandy or clay soils.

SPIKE TEPHROSIA *Tephrosia spicata*

FLOWER
(x 1)

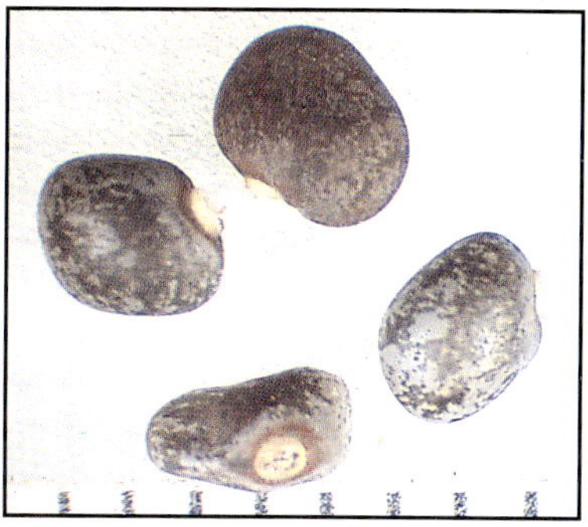

SEED

Description:
Perennial herb, with stems growing either near the ground or erect. When erect, plant supported by other vegetation. Stems reaching 2 feet. Flowers, borne on a long stalk, at first white, then changing to pink and later red. Petals falling after 3 to 4 days, so individual flowers of 2 or 3 colors and a seed pod often present at the same time on 1 plant. Blooming June to August.

Distribution:
Primarily in the Coastal Plain, southeastern United States.

Culture:
Occurs in open woodlands and along roadsides. Will stand light disking and fire. Fertilizer requirements unknown.

NARROWLEAF VETCH *Vicia dasycarpa*

PLANT

SEED

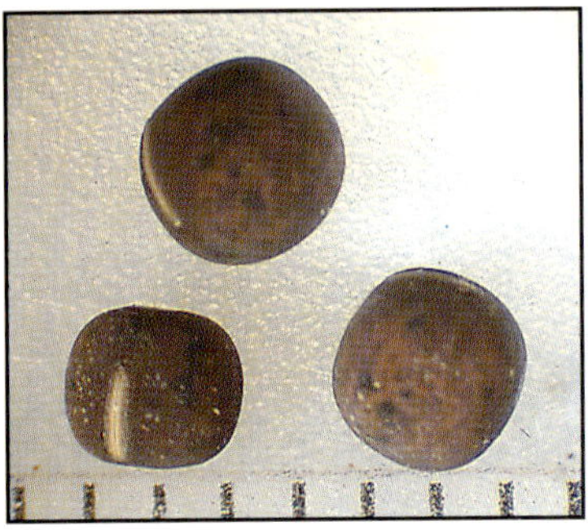

Description:
Annual herb. Stems may creep or grow erect, depending on thickness of the stand. Petals rose to purple. Blooming prolifically April to May. One of 12 species of vetch in the Southeast.

Distribution:
Southeastern United States.

Culture:
Commonly found along roadsides, in fields and waste places. It will reseed year after year. An important plant because it seeds in late spring. Grows in winter and early spring, so the best time to sow is in late fall.

GROUND NUT *Apios americana*

PLANT

FLOWER
(x 1 1/2)

SEED

Description:
Perennial vine. Stems herbaceous, twining, either spreading on the ground or climbing on vegetation. Flowers purple, red or brownish.

Distribution:
Eastern United States.

Culture:
Grows on well drained, fertile soils, usually of bottomlands. The root of this plant is edible.

BUTTERFLY PEA *Centrosema virginianum*

PLANT

FLOWER (x 1)

SEED

Description:
Perennial vine, with a deep root system. Stems herbaceous, often climbing on dead, woody stems or on low shrubs. Flower pale blue, very showy, more than 1 inch across. The pod beaked, reaching 4 inches in length, splitting lengthwise into halves that curl tightly as seed released. Blooming from June to October.

Distribution:
Southeastern United States.

Culture:
A preferred food for quail. This plant volunteers in open woodlands with mixed stands of pine and hardwood where the soil has been disturbed by disking or after a logging operation. Sites may be burned annually and disked lightly.

BUTTERFLY PEA *Clitoria mariana*

PLANT

FLOWER
(x 1/2)

SEED

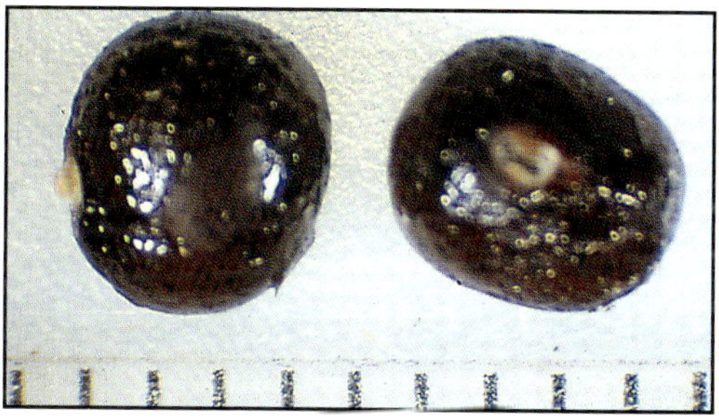

Description:
Perennial herb, with growth habits similar to *Centrosema virginianum* except stems usually do not climb. Flowers also similar in color, but distinguished from those of *Centrosema* by several distinctive features: not opening nearly as wide and the central portion being larger and more conspicuous. The pod approximately 2 inches long, much thicker than *Centrosema*. Seed sticky, not so desirable for quail. Blooming June to August.

Distribution:
Eastern United States.

Culture:
Will withstand burning and light disking. Response is similar to *Centrosema*.

DOLLAR WEED *Rhynchosia reniformis*

PLANT

SEED

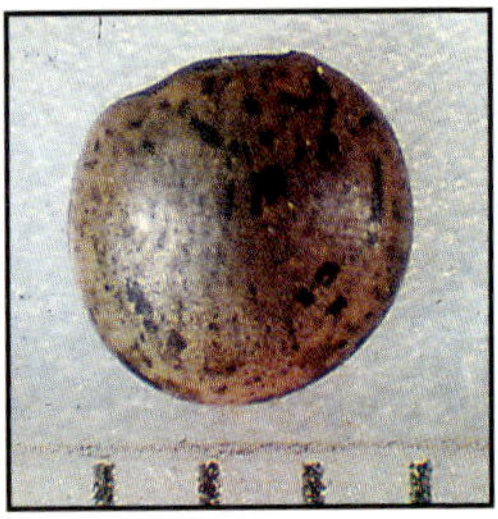

Description:
Perennial herb. Stems growing up to 9 inches high but usually only 2 to 3 inches. Leaves round, about the size of a silver dollar (hence the common name). Yellow flowers in a cluster at top of plant. Pods with 1 or 2 seed. Blooming from June to August.

Distribution:
Southeast, primarily in the Coastal Plain.

Culture:
In dry, open woods and clearings, usually on relatively exposed soil.

HAIRY RHYNCHOSIA *Rhynchosia tomentosa*

PLANT

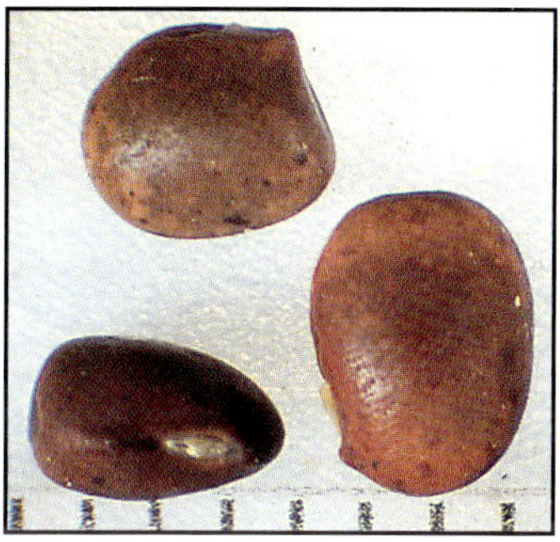

SEED

Description:
Perennial herb. Stems growing erect to a height of 2 feet. Leaves hairy, with 3 leaflets. Flowers yellow, numerous along the stem. Pods usually producing 2 seed each. Blooming June to August.

Distribution:
Primarily in the southeastern Coastal Plain.

Culture:
Grows in open woodlands and along roadsides. Will withstand fire and light disking.

WILD PEA *Vigna luteola*

FLOWER (x 1 1/4)

SEED

Description:
Perennial herb. Stems erect or low growing. Flowers bright yellow. Blooming June-November.

Distribution:
Gulf Coast, especially in Florida.

Culture:
Roadsides, in low fields and marshes. A prolific seeder that should be a desirable quail food. However, studies that might prove the use of wild pea as quail food have not yet been performed, so its importance remains unknown.

MILK PEA *Strophostyles helvola*

FLOWER (x 2 1/2)

SEED

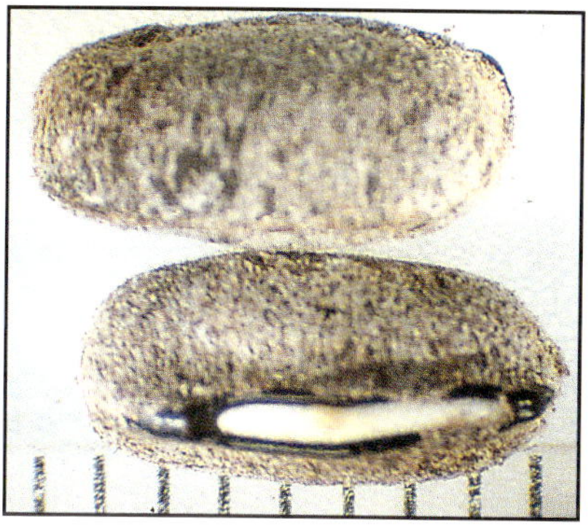

Description:
Perennial herbaceous vine, 1 to 8 feet long, commonly growing close to the ground. Flower stalks erect, 2 inches to about 1 foot long. Petals pink to purplish, with the center of the flower twisted both in this milk pea and the other, *Strophostyles umbellata.* Blooming from June to September.

Distribution:
Primarily in the Coastal Plain and Piedmont.

Culture:
This plant has a fleshy root system and can be destroyed by severe disking. It responds well to light disking, and fire will not destroy this plant. Responds to a light application of phosphate. Found in open woods and clearings. May volunteer after a logging operation.

TRAILING WILD BEAN *Strophostyles umbellata*

PLANT

FLOWER
(x 3)

SEED

Description:
Perennial herb, with stems growing along the ground. Flower stalks erect, approximately 6 inches high. Flowers pink to light red, with the center part of the flower twisted. Blooming from June to September.

Distribution:
Southeastern United States.

Culture:
Open woods and fields on sandy and clay soils. Will respond to light disking. A preferred food for quail.

KUDZU *Pueraria lobata*

FLOWER (x 3)

SEED

Description:
Perennial semi-woody vine. Flowers with a distinctive aroma similar to grape soda pop. Blooming July-August.

Distribution:
Southeastern United States.

Culture:
Planted extensively by Soil Conservation Service in the 1930s and 1940s for erosion control and hay. Can be burned and disked and responds to applications of phosphate. Not a prolific seeder.

HOG PEANUT *Amphicarpa bracteata*

FLOWER (x 3)

SEED

Description:
Annual herbaceous vine, climbing by twining. Flowers inconspicuous, those forming above ground producing several-seeded pods. Petals white to purple. Blooming from mid-summer to September.

Distribution:
Southeast.

Culture:
Found in thickets and open woodlands. Will withstand light burning.

ERECT MILK PEA *Galactia volubilis*

PLANT

FLOWER (x 2)

SEED

Description:
Perennial herbaceous vine with stems twining 3 to 4 feet high. Flowers pink. Blooming June to September. There are 6 species of *Galactia* in the Southeast.

Distribution:
Southeastern United States.

Culture:
Responds to disking and fertilizing. Fire does not damage this plant. Does not grow in a dense stand but grows best where there is little competition from other plants. Seed fall on bare soil and are sought after by quail.

WOOD SORREL FAMILY
Oxalidaceae

YELLOW WOOD SORREL *Oxalis stricta*

PLANT

FLOWER (x 2 1/2)

SEED

Description:
Annual, low-growing, erect herb. Petals yellow or green. Fruit a small, many-seeded capsule, shaped like a small okra pod. Blooming from mid-summer to frost. One of approximately 6 species of yellow wood sorrel in the Southeast; several varieties recognized in some of these species.

Distribution:
Southeast.

Culture:
Woodlands, field edges and cultivated land. Will withstand cultivation. Grows best on fertile soil.

GERANIUM FAMILY
Geraniaceae

WILD GERANIUM *Geranium carolinianum*

FLOWER (x 2)

SEED

Description:
Also called cranesbill. Annual herb, often persisting through the winter in the Deep South. Stems erect, 6 inches to 2-3 feet high. Flowers pale pink. Seed with uneven surface and green embryo. Blooming from early spring to late frost.

Distribution:
Southeastern United States.

Culture:
Found in fields and open woodlands, particularly on land that has been disturbed by cultivation or disking. Common in crops and gardens. Grows best on fertile soil. An ideal quail food because of its early fruiting in spring during critical periods of scarcity of other seed.

SPURGE FAMILY
Euphorbiaceae

DOVE WEED *Croton glandulosus*

PLANT

BRANCH

SEED

Description:
Also called tropic croton. Annual, erect herb from a taproot. Stems 1 to 3 feet high, many-branched. Flowers small, yellow. Capsule 3-lobed, with 3 seed. Blooms early summer to frost. Five species of croton found in the Southeast.

Distribution:
Common in the Coastal Plain and Piedmont but never in the mountains.

Culture:
Found in cultivated fields and pastures in most soil types. Disking encourages this plant. Not grazed by livestock. Important for quail and dove because seed fall on bare soil and are readily found. Seed from this species growing in pastures in Texas are a particularly important quail and dove food.

WOOLLY CROTON *Croton capitatus*

FLOWER (x 1 1/4)

SEED

Description:
Annual erect herb, growing to 3 feet tall. Entire plant fuzzy. Producing 3 seed per capsule. Blooming from mid-summer to frost.

Distribution:
Southeast except the Florida peninsula.

Culture:
Grows in waste places and overgrazed pastures on sandy and clay soils. Disking encourages this plant. Will respond to application of complete fertilizer. Seed are desirable quail and dove food.

THREE-SEEDED MERCURY *Acalypha gracilens*

PLANT

FLOWER (x 2)

SEED

Description:
Annual erect herb, growing from 6 inches to 3 feet high. Flowers small, yellow, along the stem at bases of leaves. Blooming from mid-summer to frost. One of 4 relatively common species of three-seeded mercury in the Southeast.

Distribution:
Southeastern United States.

Culture:
Found in fields and idle land on fertile soil. Plant encouraged by disking. In a field planted to crops, it utilizes the fertilizer residues.

FLOWERING SPURGE *Euphorbia corollata*

PLANT

FLOWER (x 2)

SEED

Description:
Herbaceous perennial from a deep taproot. Leaves alternate, appearing whorled at the base of the primary branches. Stems often much-branched above with leaves in pairs at branch points and alternate in between. Flowers terminal, in whitish clusters. Fruit a 3-seeded capsule. Blooming July-August.

Distribution:
Southeast.

Culture:
Grows in idle land, open areas, in various soil types. Stands can be burned in winter, but disking destroys the old root system and at the same time promotes seed germination.

CASHEW FAMILY
Anacardiaceae

POISON IVY *Rhus radicans*

PLANT

FRUIT

SEED

Description:
Woody vine, climbing to heights up to 50 feet. Distinguished from other woody vines by leaves with 3 leaflets. Flowers yellowish green, small, 5-petaled. Fruit small, seedlike, white to gray, round, smooth with a gray stripe. Blooming April-May. Another closely related species, poison oak (*R. toxicodendron*), grows along the ground and spreads from underground roots.

Distribution:
Eastern United States.

Culture:
An allergenic poisonous weed found in woods, thickets and idle areas.

DWARF or WINGED SUMAC *Rhus copallina*

PLANT

BRANCH

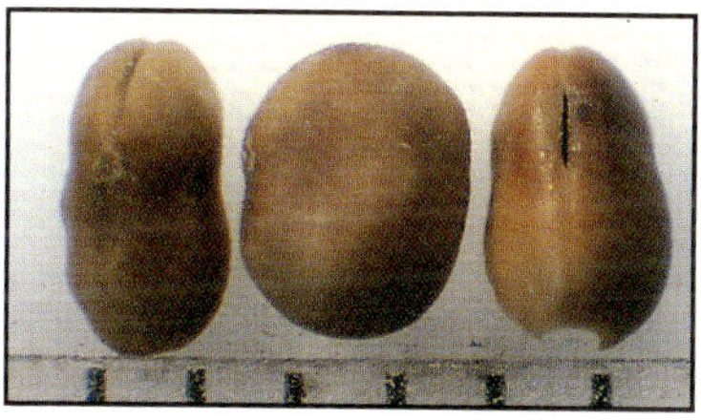

SEED

Description:
Deciduous shrub or small tree, 2 to 10 feet tall, spreading by underground stems, stems pubescent. Leaves pinnate, with 10-20 opposite leaflets. Foliage becoming a brilliant red in the fall. Petals 5, greenish white or yellow. Fruit small, seedlike, dark red-brown, in dense clusters at stem tips. Blooming July-September.

Distribution:
Throughout the United States.

Culture:
Found in open areas in direct sunlight, along roadways, fields and in open woodlands, in wide variety of soil types. Fire and disking not recommended. Quail also use this shrub in the daytime for roosting and resting.

SMOOTH or COMMON SUMAC *Rhus glabra*

PLANT

BRANCH

SEED

Description:
Same as winged sumac except that stems of this species are smooth rather than hairy. Blooming May-July.

Distribution:
Eastern United States.

Culture:
Same as winged sumac.

HOLLY FAMILY
Aquifoliaceae

HOLLIES *Ilex* spp.

There are 6 common and several uncommon species of native hollies that grow as shrubs or small trees. The common hollies have evergreen leaves except *I. decidua* and *I. verticillata.* Dropping of older leaves in fall sometimes exposes the ripe fruits. Flowers white, often borne above leaves of the previous year. Fruit red, orange, yellow or black.

HOLLY *Ilex opaca*

PLANT

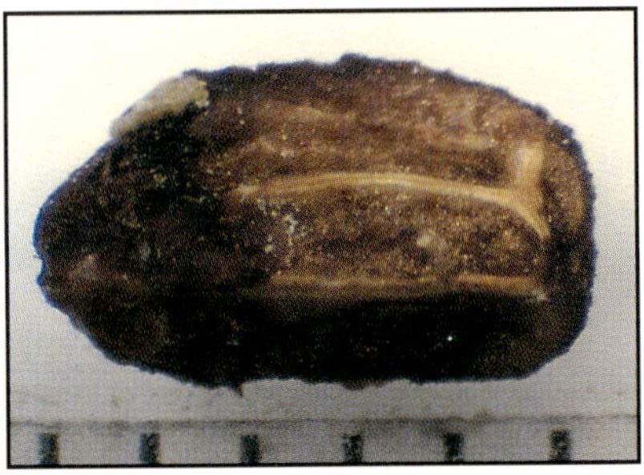

SEED

Description:
Small to medium evergreen tree with a deep root system, often with a conical crown. Leaves leathery, with few to many sharp spines around the edge. Flowers white. Fruit (drupes) round, red, orange or, in rare instances, yellow. Blooming April-June.

Distribution:
Southeast.

Culture:
Grows in a wide variety of situations and soils, most commonly in moist bottomlands. Will not withstand fire or disking. Responds well to fertilizer.

YAUPON *Ilex vomitoria*

BRANCH

SEED

Description:
Shrub, growing to 20 feet high. Fruiting in fall and persisting into winter. Blooming early summer.

Distribution:
Southeast.

Culture:
Found in all soil types. Should be protected from fire and disking. May form a dense stand.

POSSUM HAW *Ilex decidua*

BRANCH

SEED

Description:
Shrub or small tree, growing to a height of 30 feet. Dropping leaves in fall, exposing red fruit. Fruiting fall and winter. Blooming mid-summer.

Distribution:
Southeast.

Culture:
Grows best on moist soil. Should be protected from fire and disking.

MAPLE FAMILY
Aceraceae

BOX ELDER *Acer negundo*

BRANCH

SEED

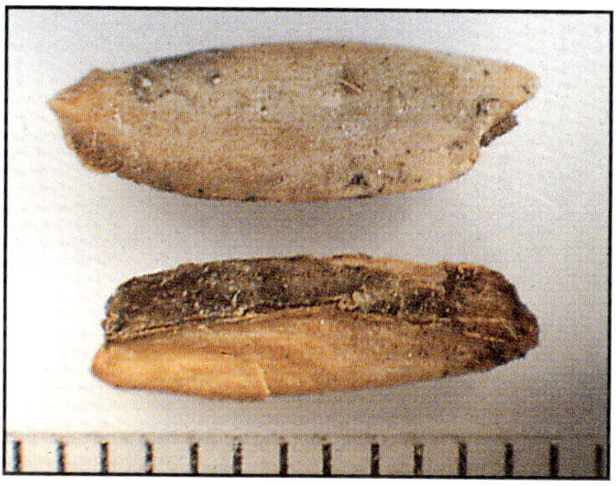

Description:
Deciduuus tree, growing to 30 feet. Flowers yellow-green, in drooping clusters. Fruit winged. Blooming early summer.

Distribution:
Eastern United States.

Culture:
Occurs on moist, fertile woodland soil, riverbanks, etc. Should be protected from fire.

RED MAPLE *Acer rubrum*

PLANT

BRANCH

SEED

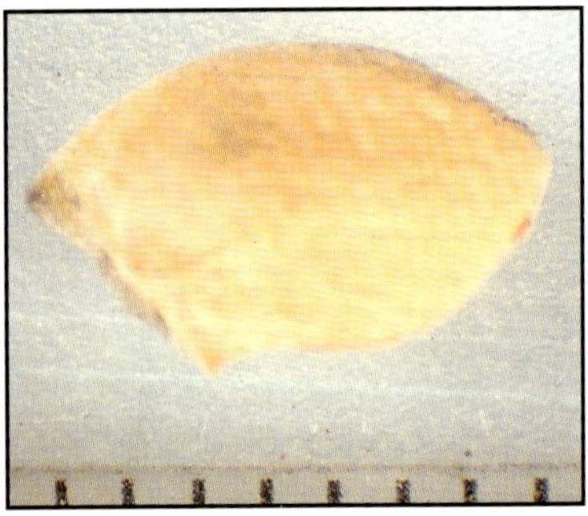

Description:
Deciduous tree, growing to 50 feet tall. Flowers scarlet, crimson or sometimes yellowish. Fruit winged, also reddish. Blooming early spring before leaves appear.

Distribution:
Eastern United States.

Culture:
Along stream banks and in wet areas, rarely uplands. Protect from fire.

BALSAM FAMILY
Balsaminaceae

JEWEL-WEED *Impatiens capensis*

PLANT

FLOWER
(x 1)

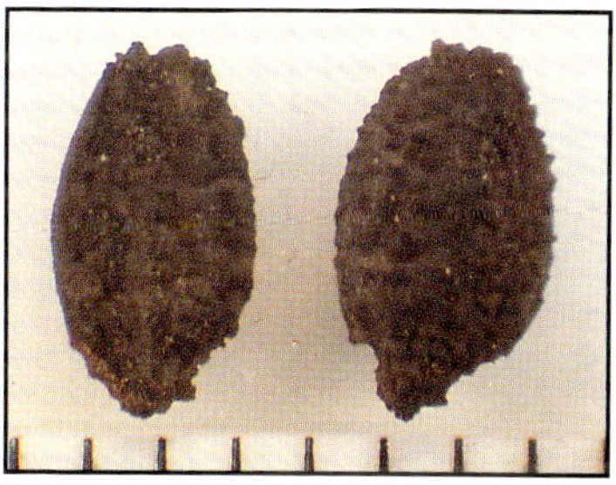

SEED

Description:
Also called touch-me-not. Annual herb from a taproot. Stems 2 to 6 feet tall. Flowers showy, orange to red. Blooming summer to frost.

Distribution:
Southeastern United States.

Culture:
Along stream margins and moist bottomlands.

BUCKTHORN FAMILY
Rhamnaceae

NEW JERSEY TEA *Ceanothus americanus*

BRANCH

FLOWER
(x 3)

SEED

Description:
Deciduous shrub, growing to 3 feet tall. Flowers small, white, in dense clusters. Blooming mid-summer.

Distribution:
Eastern United States.

Culture:
In open woodlands and field edges.

GRAPE FAMILY
Vitaceae

VIRGINIA CREEPER *Parthenocissus quinquefolia*

PLANT

SEED

Description:
Deciduous woody vine. Stems climbing to a height of 50 feet by adhesive disks that stick to vertical surfaces. Leaves palmate, with 5 leaflets. Flowers greenish yellow. Seed brown. Blooming May-July.

Distribution:
Most of the United States, except arid portions.

Culture:
Broad requirements for growth.

MUSCADINE GRAPE *Vitis rotundifolia*

PLANT

SEED

Description:
Deciduous woody vine, growing to a height of 50 feet by tendrils that attach to adjoining support. Flowers creamy white. Fruiting late summer to fall. Blooming early summer. *Vitis rotundifolia* is the most common of 5 common species of wild grapes.

Distribution:
Various wild grapes found throughout eastern United States as far west as Texas.

Culture:
Found on both damp and dry soils with high or low fertility. Will not tolerate fire. Must be pruned in early spring to avoid loss of sap. Fruit edible and widely used for jams and jellies.

MALLOW FAMILY
Malvaceae

ARROWLEAF MALLOW/SIDA *Sida rhombifolia*

PLANT

FLOWER
(x 1 1/2)

SEED

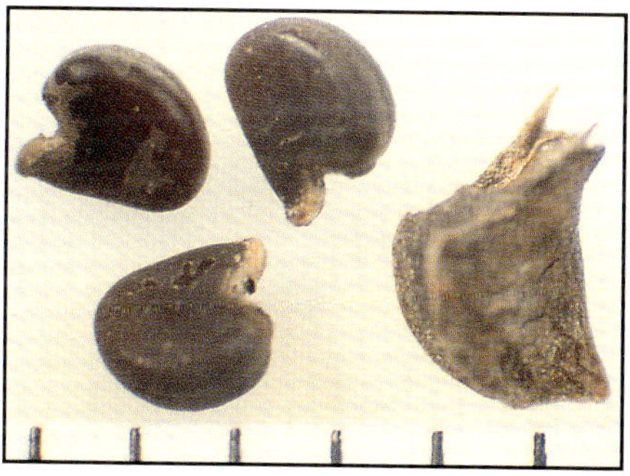

Description:
Annual herb from a tough, well developed taproot system. Stems erect, to 2½ feet tall. Flowers yellowish. Blooming mid-summer. One of 4 species.

Distribution:
Eastern United States.

Culture:
Found in waste places, around old home sites and in gardens. Grows on soil of good to excellent fertility. Will volunteer year after year by disking. A pest in gardens.

STERCULIA FAMILY
Sterculiaceae

CHOCOLATE-WEED *Melochia corchorifolia*

PLANT

BRANCH

SEED

Description:
Annual erect herb. Stems growing to a height of 5 feet, well branched. Flowers white, inconspicuous, in compact clusters at ends of stems. Blooming mid- to late summer.

Distribution:
Lower Coastal Plain.

Culture:
Grows on sandy and muck soils with poor internal drainage. Will withstand fire. Disking recommended in late spring to improve stands. An excellent quail food but distribution limited because of restrictive growth habits.

HYPERICUM FAMILY
Hypericaceae

ST. ANDREW'S CROSS *Hypericum hypericoides*

FLOWER (x 2)

SEED

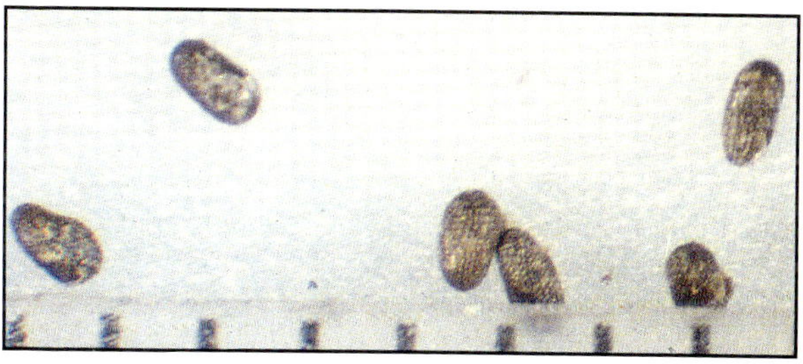

Description:
Shrub, low growing, erect. Flowers pale yellow. Petals forming a cross, resembling an "x" lying on its side. Blooming mid-summer to frost. This is 1 of about 30 *Hypericum* spp. in the Southeast, including a dozen or more that are herbaceous.

Distribution:
Eastern United States.

Culture:
Field edges in soils of high or low fertility. Fire will destroy the emergent plant, and disking will destroy entire plant.

VIOLET FAMILY
Violaceae

VIOLETS *Violet* spp.

PLANT

SEED

Description:
Annual or perennial herbs. Above-ground stems erect up to 1 foot tall, or absent. Plant lacking aerial stems appearing as a cluster of leaves and flowers. Flowers white to purple. Blooming spring to late summer. There are at least 30 species of violet in the eastern United States.

Distribution:
Eastern United States.

Culture:
Open woodlands or abandoned fields. Will grow in a wide variety of soil types with high or low fertility. Will tolerate light burning. Disking will destroy mature plants but encourages reseeding.

PASSION-FLOWER FAMILY
Passifloraceae

MAYPOP *Passiflora incarnata*

PLANT

SEED

Description:
Also called passion-flower. Perennial herbaceous vine from a deep, extensive root system. Stems low growing, up to 6 feet long but usually under 3 feet, or climbing by tendrils on adjacent vegetation. Flowers showy, white, lavender and purple. Fruit a fleshy, green, oblong berry containing many seed. Immature and mature fruits often popping loudly when stepped on, hence the name maypop. Blooming mid-summer to fall.

Distribution:
Throughout Southeast.

Culture:
In thickets and open, abandoned fields. Requires soil of moderate fertility. Will withstand fire and disking.

EVENING PRIMROSE FAMILY
Onagraceae

EVENING PRIMROSE *Oenothera biennis*

PLANT

SEED

Description:
Biennial herb, from a long taproot. Stems erect, to 4 feet tall. Flowers showy, yellow. Pods upright, opening from apex downward, with many seed remaining into winter. Blooming mid-summer to early fall.

Distribution:
Eastern United States.

Culture:
Open places along roadsides and field edges. Various soil types from poor to fertile. Will withstand fire, but excessive disking will destroy roots.

FIREWEED *Epilobium angustifolium*

PLANT

FLOWER (x 1)

SEED

Description:
Perennial herb. Stems erect, many branched, 3 feet tall. Flowers terminal, red, white and yellow. Seed 3-angled. Blooming mid-summer to fall.

Distribution:
Southeast.

Culture:
Open fields and along roadsides on sandy or clay soils. Fertilizer requirements unknown.

TUPELO FAMILY
Nyssaceae

BLACK GUM *Nyssa sylvatica*

BRANCH

SEED

Description:
Deciduous tree, growing to 50 feet tall. Lower part of trunk usually enlarged. Wood close-grained, difficult to split. Fruit a drupe, blackish blue. Blooming spring.

Distribution:
Eastern United States.

Culture:
Found on dry to wet soils. Excessive burning will leave scars on trunk.

DOGWOOD FAMILY
Cornaceae

FLOWERING DOGWOOD *Cornus florida*

BRANCH

FLOWER (x 1/2)

SEED

Description:
Deciduous small tree, growing erect to 45 feet high. Root system shallow, spreading. Leaves elliptic, turning reddish in fall. Flower clusters with 4 white bracts that resemble petals, each notched on outside edge. Fruit a drupe, bright red. Blooming early spring just as the leaves appear.

Distribution:
Eastern United States.

Culture:
In open woodlands, field edges in various soil types. Older trees will withstand fire. Deep disking close to trunk will harm root system. Widely used as an ornamental.

HEATH FAMILY
Ericaceae

SQUAW-HUCKLEBERRY *Vaccinium stamineum*

PLANT

FLOWER (x 1)

SEED

Description:
Deciduous shrub, to 3 feet tall. Flowers white, in clusters. Fruit a berry, bronze to greenish purple. Blooming mid-summer to fall. About 15 species of *Vaccinium* (blueberries) and 5 of *Gaylussacia* (huckleberries), which appear closely similar, occur in the Southeast.

Distribution:
Southeastern United States.

Culture:
In open woodlands on rocky soil. Will not tolerate fire or disking.

OLIVE FAMILY
Oleaceae

AMERICAN/WHITE ASH *Fraxinus americana*

BRANCH

LEAF

SEED

Description:
Large deciduous tree. Flowers in a loose cluster, appearing before leaves. Fruit winged, usually 1-seeded. Blooming late spring. Several other ash species and varieties also occur in the Southeast.

Distribution:
Eastern United States.

Culture:
Rich soil in moist woods. Fire will scar this tree and damage trunk. A tree of commercial value.

CHINESE PRIVET *Ligustrum sinense*

PLANT

BRANCH

SEED

Description:
Shrub or small tree, growing to 25 feet tall. Root system well developed. Flowers small, cream colored, in clusters at ends of stems. Fruit a drupe, blue, remaining on plant all winter. Blooming mid-summer.

Distribution:
Eastern United States.

Culture:
An introduced plant that has spread widely. Grows in various open or semi-open situations. Will grow in a dense stand and become a pest.

MORNING GLORY FAMILY
Convolvulaceae

DODDER or LOVE-VINE *Cuscuta gronovii*

PLANT

SEED

Description:
Parasitic herbaceous annual vine. Stems leafless, orange to yellow, twining on green plants. Flowers white. Blooming mid-summer to fall. About half a dozen species of *Cuscuta* occur in the Southeast.

Distribution:
Seed of love-vine are mixed with seed of agricultural crops; therefore this plant has wide distribution throughout the United States.

Culture:
Found growing on agricultural crops. Seed germinate on ground, then young shoots lengthen and attach to host plant, soon becoming entirely parasitic.

BUNCH MORNING GLORY *Jacquemontia tamnifolia*

PLANT

FLOWER (x 1)

SEED

Description:
Annual herbaceous twining vine, usually growing along the ground, but sometimes climbing to 10 feet. Flowers blue, borne in a dense cluster. Blooming July-September.

Distribution:
Southeast, primarily the Coastal Plain.

Culture:
In cultivated and idle fields, roadsides, commonly growing on sandy soil. Disking in winter will improve stand.

MORNING GLORY *Ipomoea purpurea*

PLANT

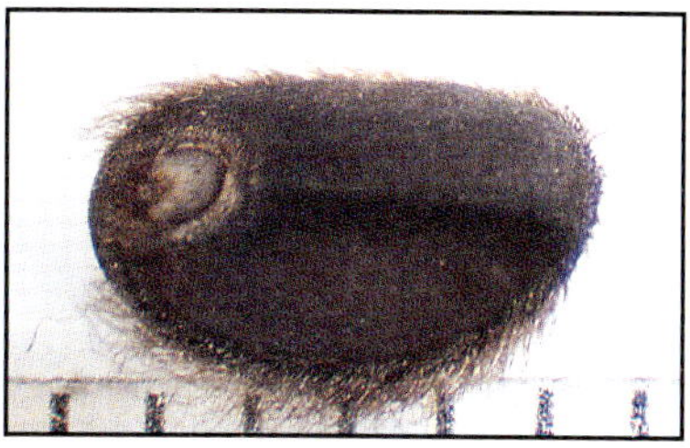

SEED

Description:
Annual herbaceous vine. Stems twining on supporting vegetation to 4 feet above ground. Flowers showy, white to purple. Blooming summer to fall. A dozen species of this genus occur in the Southeast.

Distribution:
Eastern United States and states bordering the Pacific Ocean.

Culture:
Waste places, field edges and cultivated land. Grows in all types of soil. Disking improves stand.

VERVAIN FAMILY
Verbenaceae

BEAUTY-BERRY *Callicarpa americana*

PLANT

BRANCH

SEED

Description:
Also called French mulberry. Deciduous shrub. Stems erect to 4 feet tall. Flowers pinkish, in clusters. Fruit drupaceous, 4-seeded, purple. Blooming mid-summer.

Distribution:
Throughout Southeast; not common in the Piedmont or mountains.

Culture:
In open woodlands on loamy to rocky soil. Fire will kill growth above ground.

MINT FAMILY
Lamiaceae

BLUE CURLS *Trichostema dichotomum*

PLANT

FLOWER (x 4)

SEED

Description:
Herbaceous annual. Stems erect, to 2 feet. Flowers small, showy, blue, numerous on each plant. Flower blooming for 1 day, then dropping off. Blooming mid-summer to fall.

Distribution:
Eastern United States.

Culture:
Found in open, sandy woodlands. Stands can be improved by light disking in early spring. A desirable quail food because seed usually fall on exposed soil.

SKULLCAP *Scutellaria integrifolia*

PLANT

FLOWER
(x 2)

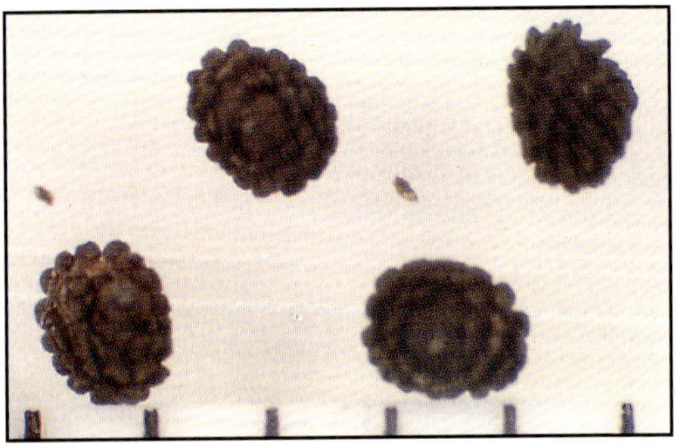

SEED

Description:
Perennial herb. Stems clumped, erect to 3 feet high. Flowers showy, blue to violet. Blooming mid-summer. About 10 other species of skullcap occur in the Southeast.

Distribution:
Eastern United States; common near the coast.

Culture:
In open areas, mainly on damp soil.

NIGHTSHADE FAMILY
Solanaceae

GROUND CHERRY *Physalis virginiana*

PLANT

BRANCH

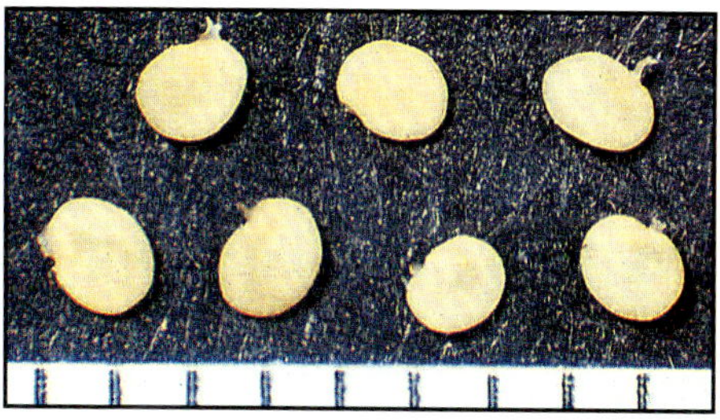

SEED

Description:
Herbaceous perennial. Stems well branched, to 2 feet high. Flowers pale yellow. Fruit a many-seeded berry, enclosed by enlarged floral parts resembling a Chinese lantern. Blooming mid-summer to late fall.

Distribution:
Eastern United States.

Culture:
Open areas. Grows in a wide variety of soils from dry to moist, loamy to gravelly.

BIGNONIA FAMILY
Bignoniaceae

TRUMPET/COW-ITCH VINE *Campsis radicans*

PLANT

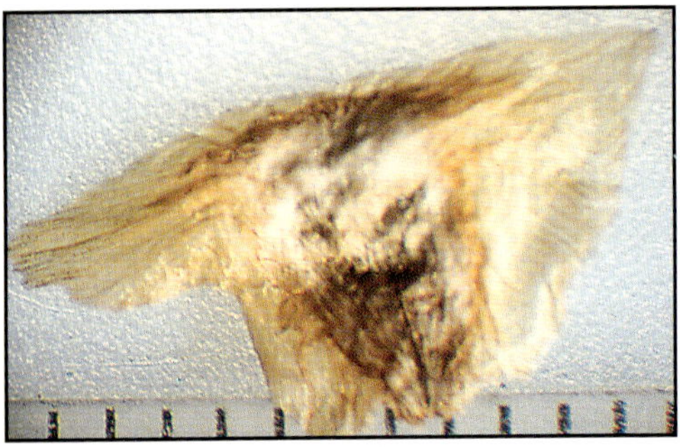

SEED

Description:
Deciduous woody vine. Stems trailing or climbing by aerial roots to a height of 50 feet. Flowers large, showy, red or orange. Seed capsule cigar-shaped, 6 to 10 inches long. Blooming summer to fall.

Distribution:
Eastern United States.

Culture:
Found along fence rows, roadsides, in woodlands and waste places. Grows in loam, clay or sand. Flowers are sought by hummingbirds.

MADDER FAMILY
Rubiaceae

POOR-JOE *Diodia teres*

FLOWER (x 2 1/2)

SEED

Description:
Annual herb, sometimes persisting through winter in the Deep South. Stems erect, many-branched, to 1 foot tall. Flowers pink. Blooming summer to frost.

Distribution:
Eastern United States; very common in Coastal Plain.

Culture:
One of the first plants to volunteer on land that has been disturbed by disking. Abundant on sandy soils and grows well under a wide range of soil fertility.

FLORIDA PURSLANE *Richardia scabra*

BRANCH

FLOWER (x 2)

SEED

Description:
Annual herb. Stems many-branched, to a height of 1 foot. Flowers white. Blooming June to frost.

Distribution:
Eastern United States, primarily the Coastal Plain.

Culture:
Found in fields and waste places. Not as common as *Diodia*. Disking encourages this plant.

HONEYSUCKLE FAMILY
Caprifoliaceae

JAPANESE HONEYSUCKLE *Lonicera japonica*

FLOWER
(x 1)

SEED

Description:
Woody vinc, trailing ur climbing to a height of 50 feet. Leaves evergreen. Flowers white to yellow, fragrant. Blooming mid-summer to frost.

Distribution:
Eastern United States.

Culture:
Introduced and naturalized in woodlands, open areas and on fences. Grows in a wide variety of soil types. Will withstand fire and light disking.

ELDERBERRY *Sambucus canadensis*

BRANCH

SEED

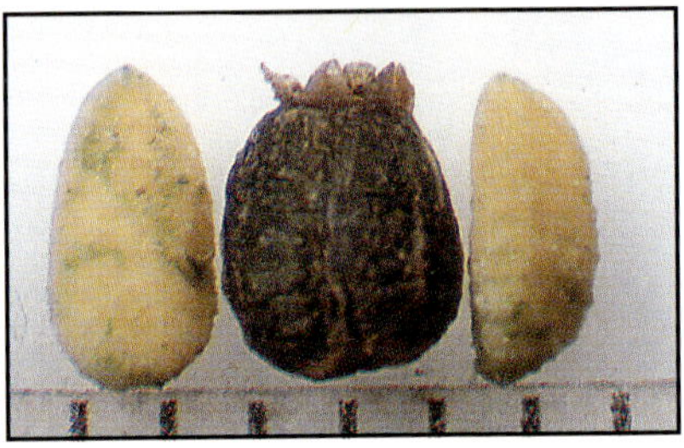

Description:
Shrub, with several to many stems commonly growing from a root crown to heights up to 8 feet. Flowers small, white, borne in a large, flat terminal cluster. Berry purple. Blooming July-August.

Distribution:
Southeast.

Culture:
Grows along roadsides, in fencerows, in open areas, especially following disturbance, usually on damp soils. Fire and disking not recommended.

ASTER FAMILY
Asteraceae

GIANT RAGWEED *Ambrosia trifida*

PLANT

SEED

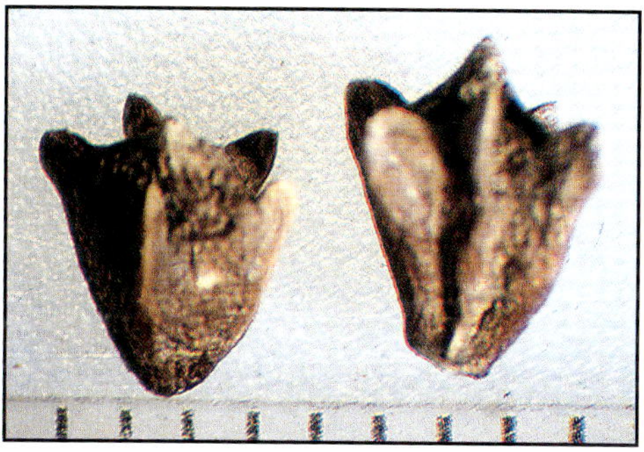

Description:
Annual herb from a large taproot. Stems erect, to a height of 10 feet. Flowers inconspicuous, mostly in long terminal spikes above leaves. Seed (nutlets) forming where leaves attach to stem and at base of pollen-producing spikes. Blooming late summer to frost.

Distribution:
Throughout eastern half of the United States, except lower Florida.

Culture:
Open areas of old pastures, woodlands, cultivated fields and along roadsides, in a wide variety of soils. Will withstand fire. Disking every few years improves stand.

COMMON RAGWEED *Ambrosia artemisiifolia*

PLANT

SEED

Description:
Annual herb. Stems erect, branched, 3 feet tall. Flowers inconspicuous. Pollen produced in terminal spikes. Seed (nutlets) forming where upper leaves attach to stems. Blooming September to frost.

Distribution:
Eastern United States.

Culture:
Found in pastures, cultivated fields, in idle lands and along roadsides. One of first plants to volunteer on moderately fertile to fertile soil after soil is disturbed. May grow in dense stands. Pollen of giant and common ragweed causes hay fever.

SUNFLOWERS *Helianthus* spp.

PLANT

SEED

Description:
Perennial and annual herbs. Stems erect, from 1 to 10 feet. Flower heads showy, yellow, sometimes with dark centers. Blooming summer to early fall. About 40 species of sunflower occur in the Southeast.

Distribution:
United States, except arid sections.

Culture:
In cultivated fields, waste places, pastures, grain fields and along roadsides in various soil types. Stands of annual sunflower can be increased by disking.

BEGGAR TICKS *Bidens* spp.

PLANT

FLOWER (x 1 1/4)

SEED

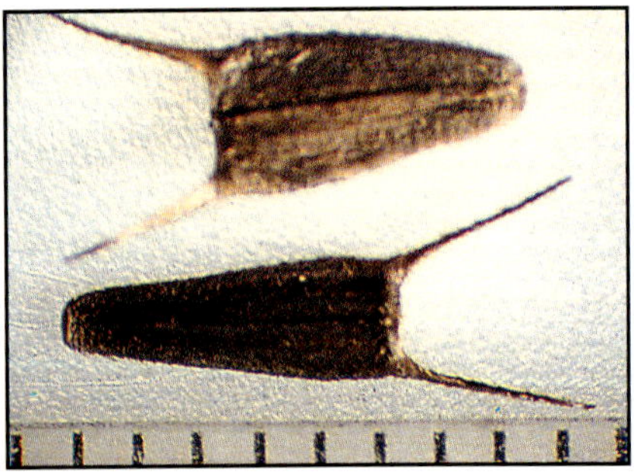

Description:
Annual and perennial herbs. Stems trailing or erect. Flowers yellow, in a disk. Seed (nutlets) with 2 pointed spines that stick to clothing and (presumably) the fur and feathers of animals. About 12 species of *Bidens* occur in the Southeast.

Distribution:
Throughout the United States.

Culture:
Found on wet or dry soils in pastures, open areas and along roadsides.

CROP PLANTS
Various species

Seed from many cultivated plants are important quail foods. Corn, field peas, soybeans and sorghum are planted specifically to produce food for bobwhites. Harvest residues from other crops are of major importance. These plants are members of families for which native and naturalized species have been introduced already. Crop species do not invade natural vegetation and rarely reach maturity if they volunteer following cultivation, so their placement under this artificial grouping seems warranted. Included here are illustrations of seed of all these plants and, in some cases, the plants that produce them. Descriptive information is given for only a few of these plants.

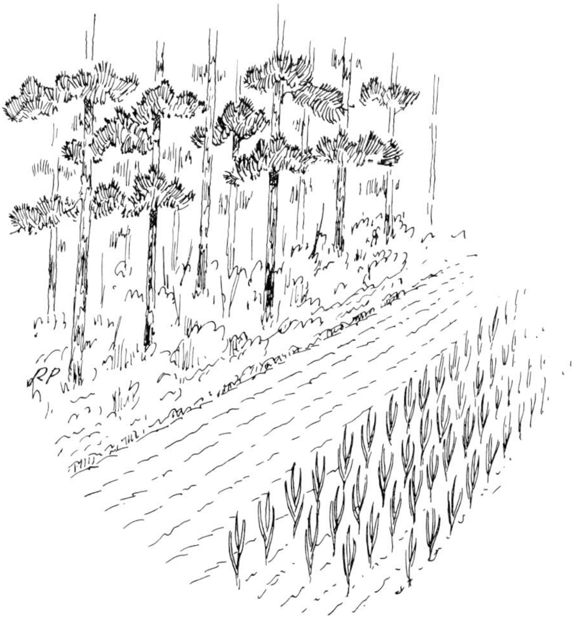

CORN *Zea mays*

WHEAT *Triticum aestivum*

SORGHUM *Sorghum vulgare*

SOYBEAN *Glycine max*

PLANT

SEED

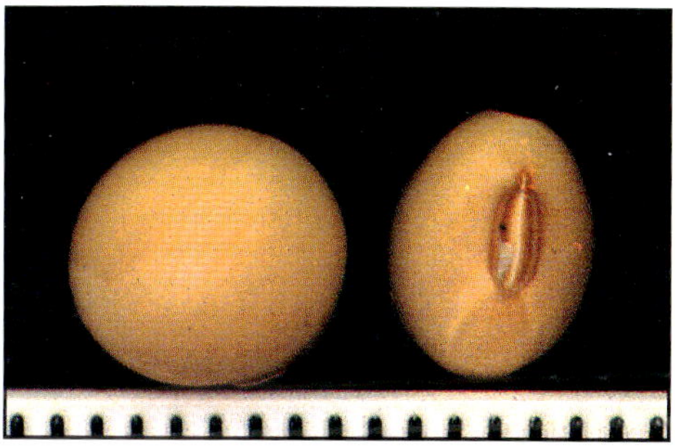

Description:
Annual herb, growing erect to a height of 3 feet. Flowers varying in color from white to deep purple. Seed exhibiting a wide range of colors and sizes among the cultivated varieties.

Distribution:
Throughout the United States, particularly in the Southeast and in California.

Culture:
Widely cultivated as a crop. Whole seed and seed fragments (caused by harvesting) frequently eaten by quail.

COWPEAS or FIELD PEAS *Vigna unguiculata*

FLOWER (x 2)

SEED

SEED

COWPEAS or FIELD PEAS *Vigna unguiculata*

PLANT

SEED

Description:
Annual herb, growing prostrate to erect up to 3 feet. Flowers white to deep purple and variegated. Seed with a wide range of colors from white to black, some variegated. A species consisting of many named varieties, or cultivars.

Distribution:
United States.

Culture:
Widely cultivated. Planted specifically for quail. Whole seed and seed fragments eaten.

PEANUT *Arachis hypogaea*

PLANT

FLOWER (x 3/4)

SEED

Description:
Annual cultivated herb. Flowers yellow, above ground, later elongating at the base of the pod (pegging) and pushing the pod underground.

Distribution:
Primarily Coastal Plain.

Culture:
Extensively cultivated as a crop; may volunteer sporadically the following year.

RICE *Oryza sativa*

BENNE or SESAME *Sesamum indicum*

RYE *Secale cereale*

BARLEY *Hordeum vulgare*

BUCKWHEAT *Fagopyrum esculentum*

GLOSSARY OF BOTANICAL TERMS

Technical terms have been kept to a minimum, but botanical accuracy is essential for a reader to make the best use of this book. This explanation of terms employed in the descriptions is included to help our readers better understand the conditions and structures described as important in distinguishing different plant species.

annual – see herb.

beak – a prolonged usually thickened projection extending from certain types of fruit.

axillary (or lateral) bud – a bud located in the upper angle between a leaf stalk and the stem, just above the leaf base. This type of bud forms the lateral branches of stems.

berry – a pulpy or fleshy fruit, usually developing showy colors at maturity and containing 1 to many seed; often used as a general category for any sweet and juicy fruit. Tomatoes and grapes are typical berries; see also drupe.

biennial – see herb.

bract – a leaf differing in size, shape, texture, color (or some other feature) from the typical green vegetative leaves of a plant and found only in association with the flowers. Poinsettia and flowering dogwood have conspicuous bracts that resemble flower parts.

capsule – a dry type of fruit, usually containing many seed, which splits open to release them at maturity. Okra pods and cotton bolls are capsules botanically.

deciduous – not persistent or evergreen; descriptive of the leaves that last for just 1 growing season in certain woody plants, then fall, usually in autumn.

drupe – a berrylike fruit with a pulpy, often colorful exterior, but the inner tissue surrounding each seed hardened into stony, 1-seeded segments called pits. Cherries and holly "berries" are drupes; very small ones, like the segments of a blackberry, are called drupelets.

elliptic – an outline or shape that resembles a flattened circle. It is usually about 2 or 3 times as long as wide, widest near the middle, and rounded at the ends.

emergent – extending above water level; descriptive of stems of certain aquatic plants in which the lower portions are either floating or rooted so that the aerial parts are supported from below water level.

endosperm – a special tissue present in mature seed of certain plants, which usually is very rich in nutrient materials such as starches and oils. Endosperm nourishes the developing plant embryo during its development within the seed prior to (and sometimes during) germination; cereal grains, the principal food source of mankind as well as many birds and other animals, produce and store large amounts of endosperm.

evergreen – persisting for 2 or more seasons; descriptive of leaves that often are tough and leathery in certain woody plants, and thus may remain on the plant for several years before they fall.

flower cluster (or head) – the pattern that results when 2 or more flowers are borne close together on a plant and collectively may form a showy display. Small flowers often are grouped into such clusters, whereas large flowers often are solitary.

flower stalk – the section of stem immediately below a flower. In many plants this stalk is the one that eventually supports a fruit.

fruit – a ripened structure, usually containing 1 or more seed, developed from the central-most part of a flower. Botanically, it is the "organ of seed dispersal" – either by rupturing or scattering seed or by virtue of surface features that enable dispersal by wind, water, or animals that either consume or passively carry them.

genus – a taxonomic group consisting of 1 or more species. A genus is distinguished by many shared morphological and reproductive characteristics among the species it includes and usually several well-marked differences with other genera to which it might be compared.

hairy – having small, fine-textured extensions from surfaces. Such hairy surfaces vary in appearance from whitish (when the hairs are very short, fine and dense) to woolly (when hairs are long and curled or much-branched) and may cause the plant to feel gummy (when small secretory glands are present on the hairs).

herb (or herbaceous) – a plant with the above-ground stem lasting no more than 1 year. Such plants are of 3 growth types: annual, growing from seed to maturity (flowering and fruiting) and then dying in 1 season; biennial, growing vegetatively during 1 season, flowering and fruiting during the second, before dying; perennial, forming annual above-ground structures (vegetative and reproductive) from an underground stem or root that persists indefinitely, often for many, many years. See also tree, shrub, and vine.

inflorescence – see flower cluster.

leaf – a structure borne by stems, usually with an expanded, flattened green portion (the blade) separated from the stem by a stalk (the petiole) and sometimes having a pair of basal appendages (the stipules) located near the stem. Blade portions of leaves may be in the form of 1 continuous area (simple leaf) or divided into separate segments or leaflets (compound leaf). Leaves are the plant organs within which the food (sugar) that maintains virtually all terrestrial life forms is made.

leaf bud – see axillary bud.

leaf stalk – a usually narrow, rather elongate portion of a leaf that extends between the leaf blade and the stem; a petiole.

leaflet – 1 segment of a leaf blade that is divided into discrete parts in a compound leaf.

lobed – a leaf type in which the blade is more or less dissected (by a deeply indented margin) into alternating lobes and sinuses (spaces). Lobed leaves resemble compound leaves and may be pinnately or palmately dissected, but the blade is 1 continuous surface rather than segments or leaflets.

ovary – the central portion of a flower, which develops into a fruit as the seed it contains mature.

palmate – a type of compound leaf in which all leaflets are joined to the tip of the leaf stalk or petiole.

perennial – living for an indefinite period that covers several to many years. All woody plants are perennial. See perennial herb.

petal – 1 of the showy, usually colorful parts in flowers. All the petals of a flower form the corolla, which enables the plant to attract insect and animal pollinators.

pinnate – a type of compound leaf in which the leaflets are arranged along the sides of a central axis (called the rachis) so that the resulting pattern somewhat resembles a feather.

pod – a general term for any type of dry fruit that splits apart to shed its seed at maturity; a capsule or legume, often used as a synonym for the latter (i.e., a bean pod).

root crown – the enlarged taproot of certain types of perennial herbs, from which the annual above-ground portions arise.

seed – botanically, this structure consists of an embryonic plant enclosed by specialized surface tissues (the seed coat) and usually provided with stored food (either as endosperm or as part of the embryo itself). As used herein, the term is also applied to 1-seeded fruits such as grains of grasses, nutlets of sedges and other plants, acorns of the oaks, and the stones (pits) of cherries and similar drupaceous fruits. The latter usage is consistent with vernacular language but falls short of botanical accuracy.

sepal – 1 of the outermost parts of a flower, usually green or greenish. Sepals enclose the other parts while the flower is in bud and collectively are known as the calyx.

shrub – a woody plant usually much-branched from ground level, thus having no main central trunk, and rarely attaining more than 12 feet in height.

species – a taxonomic group consisting of individuals that are alike in all essential respects, forms a reproducing population by sexual or asexual means, and can be distinguished from other similar groups by these morphological, physiological, and reproductive attributes. Animal species commonly are characterized by reproductive isolation in nature; plant species, for which reproductive biology and mating possibilities are usually inferred rather than observed directly, typically are defined mainly on the basis of morphological differences.

stem – the main leaf-bearing axis of the plant body, typically developed above ground level in most species. Stems may be woody or herbaceous.

stipules – basal leaf parts found in pairs near the juncture of leaf stalk and stem. Stipules vary widely in form among species: bladelike, or as tendrils, thorns, scales, etc., or may be lacking.

taproot – the main central root that initially forms upon seed germination. This root may become enlarged as a storage organ (in some biennial or perennial herbs), remain slender (in annuals), or undergo continued growth for an indefinite period (in woody plants).

tendril – a specialized stem or leaf that coils around nearby objects, thus supporting the growth of certain types of vines.

tree – a woody plant having 1 main central trunk that is 4 inches or more in diameter at 4 feet above ground level and usually grows to a height of 12 feet or more.

variety – a group of plants having certain distinctive features but considered insufficiently well marked to merit recognition in species rank. Two varieties of a plant species typically will occupy mostly separate geographical ranges and show evidence of hybridization where those ranges come into contact or overlap.

vine – a plant having a very elongate stem that climbs upon and is supported by other vegetation by some means. Vines may be woody or herbaceous and climb mostly by twining, tendrils, and adhesive roots.

REFERENCES

1. Duncan, W.H. and L.E. Foote. 1975. *Wildflowers of the Southeastern United States.* University of Georgia Press, Athens. 296 pp.

2. Fassett, N.C. 1960. *A Manual of Aquatic Plants.* University of Wisconsin Press, Madison. 405 pp.

3. Fernald, M.L. 1950. *Gray's Manual of Botany.* 8th Edition. American Book Co., New York. 1,632 pp.

4. Graham, E.H. 1941. *Legumes for Erosion Control and Wildlife.* U.S. Department of Agriculture, Misc. Publication No. 412. United States Government Printing Office, Washington. 153 pp.

5. Robinson, B.L. and M.L. Fernald. 1908. *Gray's New Manual of Botany.* 7th Edition. American Book Co., New York. 926 pp.

6. Hitchcock, A.S. 1950. *Manual of the Grasses of the United States.* 2nd Edition. Revised by A. Chase. U.S. Department of Agriculture, Misc. Publication No. 200. United States Government Printing Office, Washington. 1,051 pp.

7. Landers, J.L. and A.S. Johnson. 1976. *Bobwhite Quail Food Habits in the Southeastern United States with a Seed Key to Important Foods.* Misc. Publication No. 4. Tall Timbers Research Station, Tallahassee, Florida. 90 pp.

8. Martin, A.C. and W.D. Barkley. 1961. *Seed Identification Manual.* University of California Press, Berkeley and Los Angeles. Cambridge University Press, London, England. 221 pp.

9. Martin, A.C., H.S. Zim and A.L. Nelson. 1951. *American Wildlife and Plants, A Guide to Wildlife Food Habits.* McGraw-Hill Book Co., Inc., New York. 499 pp.

10. Radford, A.E., H.E. Ahles and C.R. Bell. 1968. *Manual of the Vascular Flora of the Carolinas.* University of North Carolina Press, Chapel Hill. 1,183 pp.

11. Sargent, C.S. 1926. *Manual of the Trees of North America.* 2nd Edition. Reprinted 1961 in 2 volumes by Dover Publications, Inc., New York. 934 pp.

12. Small, J.K. 1933. *Manual of the Southeastern Flora.* Science Press Printing Co., Lancaster, Pennsylvania. 1,554 pp.

13. United States Department of Agriculture Research Service. 1971. *Common Weeds of the United States.* Dover Publications, Inc., New York. 463 pp.

14. Van Dersal, W.R. 1939. *Native Woody Plants of the United States: Their Erosion-Control and Wildlife Values.* United States Government Printing Office, Washington. 362 pp.

INDEX

C

D

E

F

G

H

I

J

Q

R

S

T

U

V

W

Y

Z

FIELD NOTES

FIELD NOTES